MANAGING TECHNOLOGICAL INNOVATION

≡TΛ WILEY SERIES IN ENGINEERING & TECHNOLOGY MANAGEMENT

Series Editor: Dundar F. Kocaoglu, Portland State University

MANAGING TECHNOLOGICAL INNOVATION

Competitive Advantage from Change

FREDERICK BETZ

A WILEY-INTERSCIENCE PUBLICATION

JOHN WILEY & SONS, INC.
NEW YORK / CHICHESTER / WEINHEIM / BRISBANE / SINGAPORE / TORONTO

Library of Congress Cataloging-in-Publication Data
Betz, Frederick, 1937–
    Managing technological innovation / Frederick Betz.
      p. cm.—(Wiley series in engineering & technology
management)
    Includes index.
    ISBN 0-471-17380-0 (cloth : alk. paper)
    1. Technological innovation—Management. 2 Research,
Industrial—Management. I. Title. II. Series: Wiley series in
engineering and technology management.
HD45.B443 1997
658.5'14—dc21   97-2037

Printed in the United States of America

10 9 8 7 6 5 4 3 2

For Nancy

CONTENTS

PREFACE

This book is intended to provide an overview of managing change in technology, usually called management of technology (MOT), which covers

1. How technological innovation occurs
2. Concepts and techniques for managing technological innovation
3. Cases illustrating commercially successful and unsuccessful technological innovations

MOT topics continue to increase in importance, deepen in understanding, and expand in coverage. MOT is now an identifiable area of study, with a scholarly research history that extends back over 40 years.

Before 1987, the topics now included in MOT were divided among different groups of researchers who studied different aspects of innovation (Kocaoglu, 1994). One group, consisting primarily of economists, studied the macro-level of innovation process at the national scale. Their emphasis was on the fact that technological progress has made, and continues to make, important contributions to economic development and industrial productivity. In addition, there was a second group of researchers, primarily engineering management researchers, who studied the micro-level of innovation process in the industrial organization, particularly in the engineering function and research and development (R&D). Their emphasis was on how to manage the engineering function and research activities in a firm. A third group of researchers, primarily business school researchers, studied the micro-level of new high-tech businesses. Their emphasis was entrepreneurship and how to start new high-tech business ventures. Finally, there was a fourth group of researchers, primarily political scientists, sociologists, and experienced R&D administrators, who studied the macro-level of national support of research and development. Their emphasis was on how to administer governmental R&D programs and how to formulate effective science and technology policy.

By the mid-1980s, many of the different participants in these studies began to appreciate that each group was studying only one aspect of a larger topic, which focused on managing the whole process of technological innovation; they began to call for integration of a new field, which they then called the "management of technology." For example, in 1986, a group in the United States assembled in a National Science Foundation workshop to encourage the recognition of MOT as a distinct field of study. This workshop was conceived and chaired by Richie Herink (then at IBM), organized by the National Research Council of the National Academy of Science, and resulted in the publication of a brief pamphlet titled "Management of Technology: The New Challenge" (National Research Council, 1987). This group emphasized that the challenge of MOT consisted of integrating knowledge about technological innovation in the interface between the disciplines of engineering and management.

Their view of MOT could be classified by two conceptual dimensions:

- A *management* focus: strategic or operational concerns of managing technology
- A *level* focus: macro-level or micro-level concerns of managing technology

Figure 1 summarizes these concepts in order of the issues of concern then in MOT. At the macro-level of strategic focus in MOT were concerns about how to formulate and implement *science and technology policy* for national economic growth. At the macro-level of operational focus were concerns about how to best manage the *innovation processes* that create national capability for technological progress. In contrast, at the micro-level of opera-

	Strategic Management Focus	Operations Management Focus
Macro-Level Focus	Science and Technology Policy	Innovation Processes
Micro-Level Focus	Technology Strategy	Managing Engineering and R&D Activities

tional focus were concerns about how to best *manage engineering and R& D activities* within a firm, and at the micro-level of strategic focus were concerns on how to formulate and implement *technology strategy* within a firm.

While this ordering of issues in MOT continues to be appropriate, new MOT concerns had been added by the mid-1990s:

- The need to manage advances in software-based *service technologies*
- The need to manage the *integration of service and manufacturing technologies*

Service technologies consist of software-based information and communication and control systems. Even in manufacturing technologies, the importance of the software component has increased dramatically, with the cost of developing the software aspects of manufactured products often dominating the cost of developing the material aspects of these products. Moreover, it is now of great importance to integrate technical progress in both the service and manufacturing technologies of a firm.

Two other new MOT challenges have arisen, from the need to:

1. Integrate organizational change with technical change; and
2. Manage with technology tools.

The impact of computer and communication technologies has even affected how we think about what constitutes effective management practices. Computer and information technology has fostered powerful new software tools for management, by means of which operational processes and procedures are controlled. Some have called this the need to "manage with technology."

These newer MOT concerns can be summarized along two dimensions:

1. Management concepts (divided between paradigm issues and integration issues)
2. Computational technologies (divided between issues of computer-based procedures and software tools).

Figure 2 summarizes what is new in the 1990s in the growing field of MOT.

In summary, MOT can be organized by the thematic areas in Figures 1 and 2. Although a presentation according to thematic areas of MOT is intellectually useful, there has emerged an alternative way to organize MOT topics—according to the educational customers for MOT. Since the mid-1980s, the number of educational programs in MOT have increased greatly. Generally, the students in the educational programs are employed either in government or as contractors to government, or in civilian sectors of indus-

	Issues in Management Paradigm	Issues in Management Integration
Computer-based Processes	Managing with Technology	Integrating Organizational and Technological Changes
Software Tools	Managing Progress in Service Technologies	Integrating Manufacturing and Service Technologies

try. From the experiences of teachers in these programs in presenting MOT material directly relevant to the students' needs, we have understood that MOT training for the two sectors requires different curriculum emphasis:

- Governmental sectors need to emphasize the management of technical programs.
- Commercial sectors need to emphasize the commercialization of technology.

Accordingly, I have grouped the thematic topics of MOT in a way to make clearer the relevant topics to either group of students. To do this, one sees that, in both government and business, technological change is managed through:

- Leadership and attention to systems issues
- Planning and implementing new technology

One can use these categories to group MOT topics as shown in Figure 3.

In government, or in contracting to government, one is principally concerned with technical systems that provide public services, such as health, defense, welfare, public transportation, or others. Technology strategy provides the technical basis for improving public systems; the design and development of these sociotechnical systems provides the system focus of the public official or government contractor. Implementation of technical strategies occurs in technical programs performed or sponsored by government officials; the management of technical systems is then administered in government programs or contracted out to private firms.

In commercial businesses of the civilian sector, the planning focus is not just on technology strategy, but also on its integration into business strategy; the system focus is the enterprise systems of the firm in its various businesses. The implementation of new technology that requires top-level strategic attention is the launching of new business ventures, whereas the im-

	Leadership Focus	System Focus
Planning Emphasis		
Government	Technology Strategy	Technology Systems
Business	Strategy Integration	Enterprise Systems
Implementation Emphasis		
Government	Research and Development	Technology Operations
Business	High-Tech Ventures	Innovation Processes

provement of products and production and services in existing businesses occurs incrementally in the product and production and service development procedures.

Accordingly, leadership in planning and implementing new technology acquires different emphases: Governmental programs are focused on broad technology strategy, whereas a business will emphasize integrating technology strategy into business strategy. Also, in government, the implementation of technology strategy occurs in programs of technological support, broadly focused on relevant science and generic-level technology. In comparison, in industry, the implementation of a new basic technology may be focused around the new business opportunities that it can create: new high-technology ventures.

Similarly, the system focus on technological change differs in emphasis between government and industry. Government technology programs focus on the generic level for the development of new technology systems. In business, the system emphasis is on integrating new technology into the development of the enterprise system. The implementation of new technology systems in government emphasizes the operation of a generic technology system, whereas in business, system implementation issues focus around procedures for innovating technology into products, production, or services.

In order to serve managers, scientists, and engineers in all sectors and yet retain these differences in issue emphasis, the presentation of MOT topics in this book has been organized along these lines.

To summarize, in education, MOT topics now can provide:

1. A course at the undergraduate level in liberal education of all students in the study of science, technology, and society;
2. A course at the graduate level in an M.B.A. curriculum to understand the role of technological innovation in management;
3. A master's degree in MOT for technical personnel, engineers, and scientists moving into management;
4. Executive education for managers, engineers, and scientists to improve their capability of managing the technological functions of business and government; and
5. A Ph.D. degree for scholars of the interactions of scientific, technological, economic, and societal changes.

Frederick Betz
1997

___1
SCOPE OF MANAGEMENT OF TECHNOLOGY

CENTRAL CONCEPTS

1. Technological imperative
2. Technological innovation
3. Impact of technological innovation on industry
4. Range of technologies in business
5. Macro- and micro-level foci of MOT
6. Definition of technology
7. Functional bases of a society
8. Complexity in technological innovation
9. Key concepts in MOT
10. Paradigm of MOT

CASE STUDIES

Invention of Xerography
Sony Innovates but Loses the VCR Market

INTRODUCTION

The grand theme of managing technological innovation is the whole story of technological change and its impact on society. Historically, this story is both dramatic and ruthless. The drama is the complete transformation of societies in the world from feudal and tribal to industrial. The ruthlessness in technological change has been its force, which no society was able to resist and which has been called the "technology imperative." For the last five hundred years, technological change has been irresistible in military conflict, in business competition, and in societal transformations.

Two technological innovations mark the beginning of our modern era: the gun and the printing press. The gun ended the ancient dominance of the feudal warrior, and the printing press secularized knowledge. This combination of the rise of the mercantile class and the secularization of knowledge are hallmarks of modern societies. From the fourteenth century through the twentieth century, the political histories of the world are stories of the struggles between nations and peoples, wherein the determining factors are superiorities made possible by new technologies.

> *The imperative in technology is that the superior technology of a competitor cannot be ignored by other competitors, except at their peril.*

In addition to technological change, scientific progress has been a force in modern history. Although the innovations of the gun and printing press were not made on a scientific basis, afterwards, all the other major new technologies in the world have been made on a scientific basis. Continuing technological progress in the world has been made possible by the *origin and growth of science.*

> *The technological imperative has been extended by science, providing continuing revolutions in the world based on the applications of scientific discoveries and understanding.*

These are the issues we will focus on in this book: change in technology and science and their impacts on economic, governmental, and societal change. Yet, although technology change has been pervasive in modern history, the problem is not what happened, but what *can* happen.

> *The practical problem is "how best to improve technology through managing innovation."*

This is a relatively new topic, which has been called the "management of technological innovation," or, in short form, "management of technology" (MOT). The breadth of MOT is:

1. To understand the patterns of change in the past due to the technical bases of human activities, and
2. To use this understanding to improve techniques for managing technical change in the future.

On the scholarly side, MOT probes some of the most complex issues in modern history:

- What is science, and how does it progress?
- What is technology, and how does it progress?
- How does scientific and technological progress interact with and affect societal change and economic development?

On the professional side, MOT addresses some of the toughest of management problems:

- How can the R&D infrastructure of a nation be administered to foster scientific and technological progress, contributing to national economic and social development?
- How can the technological function of a business be managed to foster technological innovation for competitiveness and opportunity?

In 1994, Richard M. Cyret and Praveen Kumar looked back on the development of the field of technology management:

> Forty years ago only a relatively few perspicacious observers realized the significance of technology management for our society. . . . Technology management has become increasingly important in the modern age. . . . Organizations have to be in a position to adapt to technological innovation. In order to be in a position to adapt, management must have knowledge of the kind of technological innovation that will come . . . the firm must have a (technology) strategy for searching. . . .
>
> —(Cyret and Kumar, 1994, p. 333)

MOT provides both theory and case histories of actual technological and scientific change to illustrate and test theory. In this chapter, we begin by addressing the following questions: What is the central concept in MOT? Why is it important? Who uses it? Why is managing technology so complicated?

TECHNOLOGICAL INNOVATION

The central concept of managing technological change is the idea of "technological innovation":

> *Technological innovation is the invention of new technology and the development and introduction into the marketplace of products, processes, or services based on the new technology.*

First, a new technology must be invented. Second, the new technology must be developed and embedded into new products, processes, or services. Third, these must be designed, produced, and marketed. Technological innovation covers the full spectrum, from the creation to the utilization of knowledge for economic purposes.

Invention is the creation of a functional way to do something, an idea for a new technology.

Invention is motivated by the desire to solve problems or to provide new functional capability. For example, vaccination was invented to solve the problem of smallpox plagues. The airplane was invented to provide a new capability of powered flight.

Innovation is introducing a new or improved product, process, or service into the marketplace.

Invention results in knowledge. Innovation results in commercial exploitation of knowledge in the marketplace. The economic benefit of invention occurs through innovation. Thus, the concept of "technological innovation" combines the ideas of technological invention and business innovation.

Despite the importance of the concept of technological innovation, it has not been always well understood or well managed because the concept bridges the business and technical worlds. There has been a cultural gap between managers and engineers. In the past, business schools mostly ignored the business functions of research, technology, engineering, and development. Conversely, engineering schools mostly ignored the management aspects of engineering and the generic aspects of technology. Accordingly, the education of managers, engineers, and scientists has been incomplete in acquiring a generic understanding of both management and technology. As a result, both managers and engineers have had to round out their education about managing technological innovation through experience and continuing education. Curricula in MOT provide a compact way to gain an understanding of technological innovation.

Case Study: Invention of Xerography

This case study illustrates the invention of a technology and the steps in technological innovation going from invention to a product. The historical context consists of the decades of the 1930s to 1950s, when the invention of xerography created the copying industry.

In 1935, Chester Carlson was working in the patent office of Mallory Company in the United States. His technical background included work as a carbon chemist, printer, and then as a patent lawyer. He had technical backgrounds in the chemistry of carbon (the powder of which he was to use in his invention), printing (his invention was a printing invention), and intellectual property rights (he understood the commercial value of a good basic patent).

As a patent lawyer, he was frustrated by the errors in copying a patent for public dissemination and with the trouble with making large numbers of copies, whose quality continually decreased with number. At that time, one had to type multiple copies with multiple sheets of carbon paper—with any typing error requiring an equal multiple of corrections, and the quality of the printing decreasing according to the number of copies. As a potential customer for his invention, he clearly understood the market need for the invention.

He began experimenting in the evenings and weekends with ways to create a new copying process. His idea was (1) to project the image of a typed paper onto a blank sheet of paper coated with dry carbon, (2) to temporarily hold the carbon on spaces of letters by static electrical charges induced by light, and (3) finally, by baking, to melt the ink onto the paper in the patterns of the projected letters. This would produce a quick, dry reproduction of a typed page; the process came to be called xerography.

Carlson succeeded in obtaining a crude image, thereby reducing his idea to practice. He filed for a patent. Yet, like all new inventions, it was still not commercially efficient, cost-effective, or easily usable. It required research and development. The development of a new technology usually costs a great deal of money, takes time, and requires skilled resources. All inventors face similar problems—conceiving the invention, reducing it to practice, obtaining a patent, and obtaining support for its development and commercialization.

Carlson went from company to company seeking support. He was turned down again and again. Each company that turned him down missed one of the great commercial opportunities of that decade. (That story you may have heard about the world beating a path to the door of the inventor of a better mousetrap—it is not true. A newly invented mousetrap that uses new technology is seldom capable of catching a real mouse until after much costly research and development.)

Finally, in 1942, Carlson found some people who could envision a better mousetrap when they saw one. A venturesome group at the Battelle Memorial Institute in Ohio—a nonprofit R&D organization with a wide range of technical research capabilities—agreed to work on the development of the xerography invention in return for a share of potential royalties.

Finally, the innovative pieces for Carlson began to fall into place—inventions, patents, R&D, and commercialization. In 1945, after Battelle had developed the process to be able to show its potential, the president of a small company saw it. He was Joseph Wilson, looking for new product technology for his company, Halroid. Wilson funded Battelle for the rest of the development and then commercialized the first copiers, which he called Xerox; he then changed the name of his company to Xerox.

The rest became commercial history. Xerox created a new industry in office copying, and was one of the fastest growing companies in the world in the 1950s and 1960s.

Interesting questions from examples like this one are: Why did companies first seeing the technology not have the vision to grab it? Why can some R&D outfits, like Battelle, have better technological vision than commercial companies but not commercialize things themselves? What kind of leadership qualities do innovative, risk-taking managers like Joseph Wilson possess? These are some the kinds of questions we will pursue in this book.

IMPACT OF TECHNOLOGICAL INNOVATION ON INDUSTRY

One of the impacts that innovative, risk-taking managers, such as Wilson, have had on society has been sometimes to create whole new industries.

The concept of technological innovation is important because it does affect the development of industry and economy. Its impact begins with science and technology contributing to the global industrial revolution. The impact of new technologies based on new science has created many basic new industries that fueled economic expansion. These periodic economic expansions are called "Kondratieff long waves." The high-technology industries of their time improved the functional capabilities of society and created new wealth.

A common pattern underlies the rate of progress in any new basic technology; this has been called the "technology S-curve." Technologies at first improve rapidly, then linearly, and finally level off. This pattern influences the growth of a new industry based on a radically new technology. It results in finally limiting the growth of the industry. The market for any industry eventually saturates after the rate of progress in its core technologies slows down. This pattern is called a "core-technology industrial life cycle."

At the firm level, technological innovation can be used to gain and sustain a competitive advantage against attacking competitors or to create a disruptive competitive advantage and attack another competitor. Once a new basic technology has been innovated, which starts a new industry, a large number of new firms are begun or enter this new high-tech industry. The firms that emerge as dominant are those called "first movers." To sustain a dominant position, these firms must establish strong engineering and research capabilities. When the industry becomes a mature-technology industry, these firms must have the market position and manufacturing capability to survive the inevitable "shake-out" in the industry when excess capacity occurs.

If a substitute technology is later innovated, dominant firms will face industrial restructuring. A substitute technology is called a "next-generation technology" and requires a new science base from the older technology. Because new science bases are needed, existing technological and engineering expertise in an industry becomes obsolete, along with the production capabilities of the existing firms. Markets can also change under the new technology substitution. Accordingly, the existing core competencies of a firm can become obsolete and its former market dominance challenged by new upstart firms.

RANGE OF TECHNOLOGIES IN BUSINESS

Technological innovation is also important because it is a major driving force for changes in a business. A business produces products for a customer, which the customer uses in an application. Technology is embedded in the design of the product, in the production of the product, and in the operations of the business that designs and produces the product. In discussing economic activities, it has been traditional to distinguish between goods and services as different kinds of products of firms. An economic good is a product expressed in physical form, such as an automobile, a computer, or a machine. An economic service is a product expressed as an activity that delivers benefits, such as air transportation, education, medical services, recreation, entertainment, food services, or financial services.

In the production of a hard good, such as an automobile, technologies are used in the design of the engine, fuel system, chassis, control systems, body, and running gear among others. Different technologies are used in the production of the automobile, such as casting of the metal engine block and parts, machining to finish cast parts, stamping of body parts, and assembly of parts into automobile components and systems. And different technologies are involved in transporting the assembled automobiles to dealers and in the follow-up fueling, maintenance, repair, and disposal of automobiles.

There are different technologies for service products, such as air transportation. In the design of an airplane, there are technologies of materials, jet engines, control systems, and lift and flight dynamics. In the production of an airplane, there are different manufacturing technologies for the production of parts and assembly of parts into an airplane. In the delivery of air services, there are also technologies for making reservations and selling tickets, scheduling flight operations and maintenance, operating airports, and controlling air traffic.

We will use the term "product" to indicate any product—good or service—and the term "production" to mean either the manufacture of goods or delivery of services.

A third economic concept has been to distinguish between technologies used internally for operations or for the external delivery of goods or services. Internally, information technologies are developed for communication and management. Externally, information technologies are developed for reaching the customer and delivering services and products to the customer. We will use the term "information" technologies to cover the kinds of technologies that are critical to communication, management, and delivery operations of a firm, whether internal or external.

Thus, the range of technologies that MOT deals with can be broadly categorized as

1. Product/service technologies
   = "**product technologies**";
2. Manufacturing/service-delivery technologies
   = "**production technologie**s";
3. Information/operations technologies for management control
   = "**information technologies.**"

As you can see, this range of technologies makes the management of technology essential to any kind of business.

FOCUS ON MOT: MACRO-LEVEL AND MICRO-LEVEL

MOT is used by practitioners in both government and in business. These two different uses create two different views on innovation by MOT—macro- and micro-levels. At a macro-level of innovation, the scope of the study of MOT is about (1) how scientific technology arises from science and, in turn, stimulates further scientific progress and (2) how scientific and technological progress stimulates economic development and societal change. At a micro-level of a firm, the scope of MOT is about how to create or obtain new technology for business opportunities and competitiveness.

These perspectives arise from where practitioners of innovation sit in the R&D infrastructure. They work in government, university, or industrial sectors. Governmental and university practitioners of MOT generally see technological innovation from a macro-focus, and business practitioners most often from a micro-focus.

Government uses technologies for defense, for the delivery of services, and in the management of government legislative, executive, and judiciary functions. In addition to the use of technologies, government supports the advancing and communication of the knowledge bases for technologies in its support of education and of research and development. The management of technical programs in government thus divides into how to use technology and how to support technological progress. Technology is used either as hardware in government services or as service technologies in the delivery of government services. Technology is also used as internal services in the administration of government activities. At the national governmental level, the issues of how best to advance the scientific technology capabilities of a nation have been called the science and technology (S&T) policies of a country.

In business, technological innovation pervades all the functions of any business because all utilize some technology. Traditional functions of a business include: (1) production, (2) marketing, (3) finance, (4) administration, (5) engineering, and (6) research. Production uses technologies in the production of the firm's products and/or services. Marketing, sales, and distribution use technologies in the information, communication, and transportation systems that help sell and deliver the firm's products and/or services. Finance uses technologies in the information and communication systems that assist in tracking and controlling the firm's performance. Administration uses technologies in the information and communication and training systems that assist in recruiting, supervising, and training personnel; in management information systems that use computers; and in communications systems to provide the internal services for decisions and cooperation among managers and engineers. Engineering uses technologies in the design of the firm's products and/or services. Research uses technologies in the methods of studying the science bases of technologies and for inventing and understanding new technologies.

In universities, science and technology are the basis for educational programs in scientific and engineering disciplines and in technical training. New science, engineering, and technology are also the focus of university research.

DEFINITION OF TECHNOLOGY

The concept of technological innovation is complex for several reasons. The first reason is the generic breadth of the concept itself. There are many kinds of technology, but a generic definition of technology that applies to all technologies is

Technology is the knowledge of the manipulation of nature for human purposes.

The etymology of the term "technology" is indicative of the fact that technology is a form of knowledge. The first part of the word comes from the Greek word "technos" (meaning the process for doing something). The second part, "ology," also comes from Greek, and means a systematic understanding of something. So technology is the knowledge of doing something—knowledge of a *functional technique.*

For example, automotive technology provides a functional technique of powered land transportation. This technology uses gasoline derived from nature to propel automobiles constructed from the materials of nature to transport people and goods across land. As another example, computer technologies use electricity derived from nature in material devices constructed from nature for aids to the functions of human thinking and communication.

Of course, there are many techniques that we ordinarily do not include in the modern use of the term "technology." For example, there are many techniques in the arts that we would not usually call technology, such as techniques for painting, acting, writing, sculpting, and so on. The term "nature" in the definition distinguishes technology from other arts and techniques, for technology is all the techniques that use nature for functional capability. Technology has a principal purpose of utility as opposed to other purposes, such as aesthetics.

The use of nature through technology often requires a manipulation of natural states. For example, oil created by nature from fossil remains is found, extracted, and refined into petroleum products by means of technologies. All the steps of changing initial states of nature into technology-final states of nature are yet states of nature.

Nature used by technology, in natural states or in technologically manipulated states, is still nature.

This is an important point because this is the fundamental basis that ties economy to environment. Technology neither creates nor destroys nature, but only alters nature.

Also, it means that the potential of any technology is ultimately limited by science. Technology can use only nature that has been discovered. Technology can systematically improve on how it manipulates nature only to the extent nature is understood. This is why science is essential to continuing technological progress, and this is why *the scope of MOT includes both technology and science.*

NATURAL BASES OF SOCIETY

The functional capability of manipulating nature is basic to society, because all individuals in society are biological animals, requiring the satisfaction of physical needs (such as air, water, food, clothing, shelter, and safety) to survive. Technology is the knowledge of how to obtain the things from nature needed to satisfy the physical needs of society.

Of course, there are many other needs of society other than physical ones (such as social needs for communication, justice, and education). Technology also affects the capability of society to fulfill its social needs (for example, military technology or communications technology). Accordingly, for MOT, it is useful to broaden our conception of nature to include society itself. From a biological perspective, human society can be conceived as a type of the social nature of animals. In this broad sense of nature (as both the physical and social aspects of the human animal), technology is the knowledge of the manipulation of both physical and social nature to satisfy human purposes.

*This broad way of thinking about technology is as providing knowledge bases for all the **functional capabilities** of a society.*

Capabilities of a society include business technologies, governmental technologies, and cultural technologies. The power of the technological imperative arises from this basic role of technology in society. Change technical knowledge in one society, and one can change the capabilities of that society. And this may alter that society's relative power within its fellow societies. This is why, historically, the tribes and nations that developed superior military technology ahead of their neighbors were able to use such improved military capability to conquer their neighbors. This is also, historically, why firms that innovated new products with superior technical capability have put their competitors out of business.

COMPLEXITY OF TECHNOLOGICAL INNOVATION

There are other reasons also for the complexity of the concept of technological innovation due to: (1) interactions, (2) systems, and (3) dynamics. Figure 1.1a illustrates the interactional complexity in technological innovation, overlapping sets with business, industry, university, and government sectors and product, customer, and application sets.

A Venn diagram is a symbolic way of showing the overlapping areas of logical sets of things to show logical interrelatedness. What this Venn diagram emphasizes is that there are many interactions among technology, business, industry, universities, and government and among technology and product, customer, and application.

Technology is one of the knowledge bases that a business uses, and technical knowledge is created and communicated not only in a business, but also at an industrial-sector level, in university education and research programs, and in governmental R&D support activities.

The direct connection between technology and business and customer is through the "product" embodying the technology that the business sells to the customer. Ultimately, all the technology used by a business is directly or indirectly seen by a customer through the product. Yet the way a customer evaluates a product is not the view the business sees of the product. The customer views the product from the context of the application in which the customer uses it, not from the view of producing the product as the business sees it.

Accordingly a business may not fully see the customer's application and, therefore, may fail to design the product optimally for the customer. In fact,

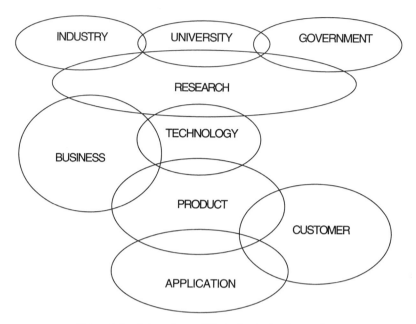

Figure 1.1*a***.** Interactions within technological innovation.

in studies on technological innovation, one of the most common reasons for product failure has been management's failure to understand the application context in which the customer evaluated the product.

Research connects the industry, university, and governmental sectors to both a business and to technology. A business uses both research and technology in the design and production of its products. It is important to note that the business and its customers are connected through the products of the business. Customers directly experience both the product and the application of the product, yet business does not directly experience the customers' application except through its product.

These connections are important for several reasons. First, the technological sophistication of a business is bounded by the research capability of the industrial, university, and governmental R&D infrastructure in which a business exists. Second, the research and technological capability of a business is known to a customer only through the business products. Research and technical capability not contributing to useful, high-performance, high-quality, and inexpensive products are not valuable to a business; therefore, businesses are very selective about what research and technologies engage their interests. Third, since the satisfaction of a customer with a product depends on its performance in an application and since a business does not directly experience the application, it is the application that is the greatest source of uncertainty about commercial success to a business in the design of its products.

Case Study:—Sony Innovates but Loses the VCR Market

This case study illustrates the logic of the difference between the product and its application. The historical setting is the early 1970s, when the consumer electronics industry continued to change through the opportunities provided by transistorized circuitry.

In the 1970s, Sony innovated the video cassette recorder (VCR) for the consumer market. The video recorder had been invented in the United States but used there only for a commercial recording market. To innovate for the consumer market, Sony had to reduce the size of the recording unit and all its parts, while improving the quality of the image resolution. This was a challenging technical problem, and Sony succeeded, creating a new market for VCRs.

Immediately, the market began to grow rapidly, and several electronics companies wanted to license Sony's technology to produce and sell VCRs. Sony refused to sell licenses, and the companies formed an industrial consortium to create their own technology. Critical to this competition was the format for recording the video image on the tape of the VCR. Sony had copyrighted its format as "Beta." The consortium created its own format and called it "VHS." There were thus two incompatible formats on the market.

At the time, the Beta format provided a better picture with a higher resolution, but this is not what the public bought. The market moved overwhelmingly to VHS format; the Beta format was eventually dropped by Sony, which lost dominance of the VCR market.

The public chose VHS over Beta because the early tapes that Sony marketed could record for only one hour. When the VHS consortium introduced their VCRs, they produced tapes of four-hour recording length. They had studied the market that Sony pioneered and had discovered that the major application for the consumer VCR was "time-swapping:" The consumer was using the VCR mostly to record a commercial broadcast that occurred at an inconvenient time to the consumer and replay it at a more convenient time. The critical product feature to the customer was the length of recording time (and not the recording quality), because movies and sporting events on TV required at least two hours' recording time.

What Sony did not correctly envision in innovating the consumer VCR was the applications of the recordings and, because of this, it did not design the right features into the VCR when it innovated the market.

TECHNOLOGY AND SYSTEMS

In addition to the logical interrelatedness between technology and the other aspects in Figure 1.1*a,* another source of complexity in technological innovation is that technology is a system, and all the things to which it relates in technological innovation are also systems. None of the systems is simple. Accordingly, we can modify Figure 1.1*a* to that of Figure 1.1*b,* to emphasize the many systems involved in technological innovation.

Industry can be seen as a system (i.e., an industrial value-adding chain). Universities are educational and research systems. Government is a system of agencies, each of which is an organizational system. Any technology is a system, which creates functional transformations. A business is also a system, involving transformations of value addition in producing and selling products. Moreover, any business system uses not just one technology system, but many technology systems. Also, in a diversified firm, there can be several different businesses, each a different system using different kinds of technology systems. Thus, the number of technology systems in a diversified firm may be very large.

A product is also a system, and embodies several technology systems. Furthermore, there is not just a customer or a set of customers for a business to sell to in a market system. The market system requires access to customers through advertising, and retailing and distribution channels for getting the product to market. The customers in the market are themselves involved in kinds of system, such as organizations and affiliations. The applications for which the product systems are used are themselves systems. And any application system uses several product systems.

In summary, complexities in technological innovation arise from the

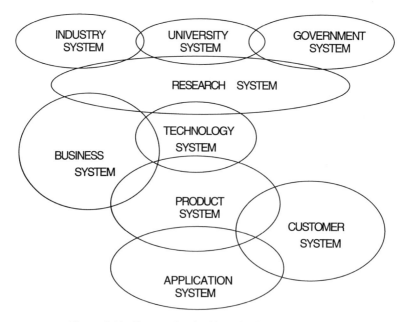

Figure 1.1*b.* Systems involved in technological innovation.

many different kinds and numbers of systems involved. But here is the fundamental problem in managing technology:

- Advances in technology come from the research system.
- Performance in a product system comes from the technology system.
- Decisions about technology and product come from the business system.
- However, commercial success comes from the customer's perception of the application system.
- Therefore, the formula for commercial success requires the proper integration of these different systems.

You can see now why MOT is not a simple subject. Moreover, complexity in technological innovation arises not only from interactions and systems, but also from change and from timing. This additional source of complexity can be added to our diagram, as illustrated in Figure 1.1c. For example, discontinuous technological change in the industrial system will create a restructuring of the industry. Technological change in the university system will contribute to scientific progress. Changes in governmental science and technology policies affect the levels and kinds of research activities a government supports.

Changes in the research system occur through individual research projects

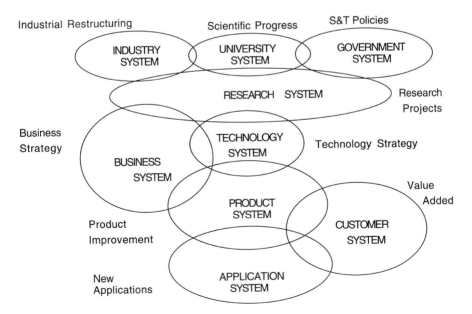

Figure 1.1c. Changes in systems involved in technological innovation.

within research programs. Technology systems used in a business are changed through technology strategy. Business strategy within the business stimulates and is impacted by technology strategy. Product improvement in performance or in production occurs from improvements in technology systems. The customer sees the product improvement as adding value to the use of products in applications. New high-tech products can create new applications for customers.

In a business, changes in technology strategy result in changes in the research system. Technological innovation can affect the market share position of a business. A product system needs to be redesigned according to technological change. Market system change occurs due to technological innovation.

What this figure emphasizes is that technological innovation affects the industrial, industry, government, business, products, market, and applications interrelations through changes in each of these. MOT has developed analytical concepts to describe many of these changes.

The problem of managing technological innovation successfully lies in dealing correctly with all the complexities arising from (1) interactions, (2) systems, and (3) change.

KEY CONCEPTS IN MOT

The concepts and techniques in MOT needed to deal with the complexity of the concept of technological innovation generates the need for a deep understanding of organizations, systems, and strategy. Accordingly, there has evolved a set of core techniques for MOT, which include:

1. Organizational analysis
2. Systems analysis
3. Technology forecasting and planning
4. Innovation procedures
5. Technical project management
6. Marketing experimentation
7. Entrepreneurship

The problem of incorporating new technologies into existing organizations requires understanding of how to redesign and restructure organizations to accommodate and exploit new technological opportunities. Systems analysis consists of techniques for defining objects as functional and connected systems. Technology forecasting and planning incorporates techniques and procedures for anticipating and planning technological change.

Innovation procedures are techniques and procedures for managing the logic of technological innovation, from invention through development to production to market introduction. Technical project management consists of procedures for managing finite, one-of-a-kind technical projects in organizations (in contrast to management techniques for managing continuing operations of an organization). Marketing experimentation employs techniques for introducing radically new high-tech products into the marketplace, creating new markets, growing new markets, establishing industrial standards for products, and so on. Entrepreneurship consists of techniques for starting new businesses or restructuring existing businesses.

MOT PARADIGM

As studies about scientific and technological progress have been accumulating and as techniques for managing technological innovation have been developed, there is now emerging an intellectual *paradigm* of MOT. In the 1980s, many who had been involved in studies about R&D management and science administration got together and emphasized that technological innovation is an intellectual topic lying in the interface between engineering and management (NRC, 1987). Since then, filling in that interface has been the focus of the MOT research and education community, which has been building a shared understanding that the interface between engineering and management is the *integration of technology systems into enterprise systems.*

This integration consists of viewing the totality of technological innovation as interactive changes between *systems of knowledge* (science and technology) and of *systems of economy* (S&T infrastructures and business systems). In this paradigm, the following kinds of issues are central to MOT:

1. Understanding long-term economic development
2. Understanding how national S&T infrastructures contribute to competitiveness
3. Forecasting change in product, production, and service technologies
4. Effectively managing the engineering and research functions in business systems
5. Integrating technology strategy into business strategy

The progress of knowledge has been, and continues to be, one of the major assets of capitalism. It was the basis for the first industrial revolution, and continues to drive the industrial revolutions in the world.

Accordingly, the paradigm of MOT is vital to modern management. Managers need to deal not only with business strategy, capital, organization, and

human resources, but also with knowledge strategy and the professionalization of the workforce. MOT influences management thinking as: *how to deal
with the advancing frontiers of knowledge in industrialized societies.*

SUMMARY

MOT is the study of the past and future of the origins and impacts of scientific and technological progress. Its central concept is technological innovation, which denotes activities spanning the invention, development, and introduction of new technology into the marketplace. Technological
innovation is complex because of different sectoral interactions, the many
systems involved, and the dynamics of creating and implementing new technology.

MOT topics can be focused at two levels of interest: (1) the macro-level
focus, at the level of a nation, or (2) the micro-level focus, at the level of a
firm. The MOT paradigm provides managers with conceptual tools to deal
with "progress in knowledge" as an industrial asset.

FOR REFLECTION

Identify an industry begun after 1850. What were the key technologies important to its establishment? What are the scientific bases of these technologies? What were the early firms that grew to dominance in the industry? Do
they still survive?

____2
INDUSTRIAL STRUCTURE

CENTRAL CONCEPTS

1. Industrial value-adding chains
2. Economic long waves
3. Factors in successful industrialization
4. Global competitiveness
5. Technology intelligence
6. Technologies of power and thinking

CASE STUDIES

Industrial Value-Adding Chain in Energy
Economic Long Waves of the Nineteenth and Twentieth Centuries
Three Hundred Years of Industrialization
Whirlpool Becomes a Global Company

INTRODUCTION

The innovation of new basic technologies has created new industrial structures. Industrial structures are organized around sectors that use the sequence of technologies needed to create the value-addedness of the functionality that connects nature to consumer. These technology-based industrial structures have been essential in the economic development of the world.

We have noted that the development of modern enterprise systems began with what has been called the "industrial revolution" throughout the world. This was a momentous event, for everyone afterward began living in a different world—one in which industrial society became a standard form in human evolution. Although the industrial revolution began in England, when we think about the industrialization of the world, we must think over a much longer period, at least three hundred years. The industrialization of the world is not over but continues into the twenty-first century. This new phase of industrialization is now being called "globalization."

We will review how continuing technological change has impacted and continues to impact the economic development of the world.

INDUSTRIAL VALUE CHAINS

The chain of industrial sectors supplying all the value-adding transformations required to get natural resources into final products for customers is called an "industrial value chain" (Porter, 1990; Steele, 1989). The general form of an industrial value chain is shown in Figure 2.1.

Industrial value chains begin with some kind of resource-acquisition industrial sector. The next sector is usually a materials-processing sector. Resource-acquisition and materials-processing sectors usually organize separately, because the technologies, investments, and operations differ markedly between acquisition and processing operations.

Materials-processing sectors usually supply materials to part-producing sectors. For example, the automobile industry is supplied by firms producing many different kinds of parts, such as tires, wheels, brakes, battery, electronics, and the like.

The next industrial sector is a major device fabricator sector. For example, in transportation applications, key major devices are automobiles for land, airplanes for air, and ships for sea. In addition, within a major device sector, subindustrial sectors are frequently organized. For example, the automobile industry divides into cars, trucks, tractors, motorcycles, tanks, and so on. The airplane industry divides into military and commercial sectors.

Sometimes, major devices are retailed directly to customers (such as cars); sometimes, they are integrated into a service-providing industry (such as airlines).

In either case, major device industries or service industries often retail through a distribution system. For example, in the case of cars, there is an automobile-dealer industry for sales and service. In the case of airlines, tickets may be sold directly by airlines or through a travel-agent industry.

Finally, the customer is served by the products and services created in the industrial value chain. These customers may be consumers, business firms (involved themselves in different industrial value chains), or government.

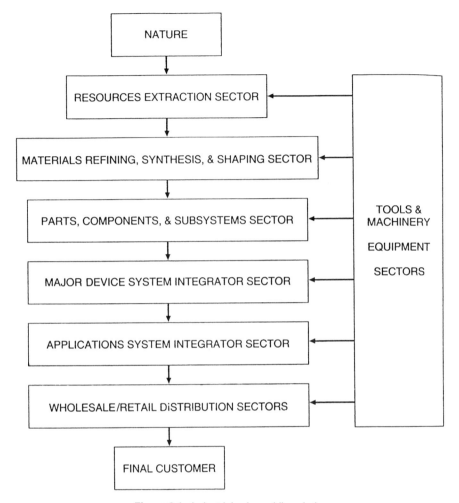

Figure 2.1. Industrial value-adding chain.

The industrial structures of an economy consist of all the transforming sets of industrial value chains that connect nature to customer.

When a new technology creates a new functionality, an industrial structure will grow to supply products to the customer embodying the new functionality.

The concept of the industrial structure provides a way to analyze the economic leveraging of the new technology. Not only are new firms established to produce the new technology, but new firms are also established to provide supplies and parts to that firm. In addition, new firms may be established to distribute and service the new product. Thus, the total market creation of a new functionality technology goes far beyond the market itself for the final product.

An industrial value chain can also be used to identify all the kinds of technologies necessary to an industrial value chain in order to satisfy customer needs from nature. Moreover, the science bases relevant to industrial value chains can then be traced from these technologies.

Case Study: Industrial Value-Adding Chain in Energy

One of the key economic resources extracted and processed from nature is energy. Modern industrial societies became based on energy extraction and use. Figure 2.2a sketches how the energy we use on earth derives from the energy/mass fusion reactions of the sun. Radiant energy from the sun powers the weather cycles on earth that transfer water in a hydrological cycle from the oceans to land and back to the oceans again. Radiant energy from the sun, along with rainfall from the hydrological cycle power the growth of biomass. Biomass, ancient and modern, provides energy sources to society in the form of coal, petroleum, gas, and wood. In addition, the hydrological cycle provides energy in the form of hydroelectric power, as rivers return water to the ocean. Wind and wave motion can also provide energy sources.

Figure 2.2b shows how an economy can use these cycles of nature to provide energy to a society. This involves a sequence (chain) of industrial sectors to acquire energy, process energy, and distribute energy to consumers in an economy.

The first industrial sectors in this sequence acquire energy for society

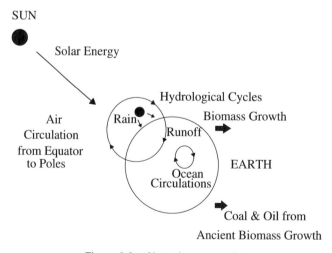

Figure 2.2a. Natural energy cycles.

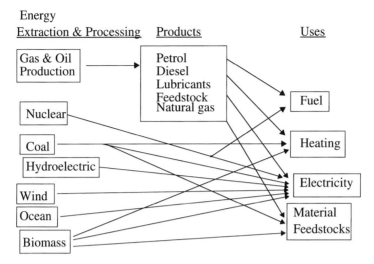

Figure 2.2b. Industrial energy sectors.

through mining of coal and uranium, production of gas and oil, logging of timber, or wind or wave generation of electricity; and thus, the first industrial sectors in the energy chain are the energy-extraction industrial sectors of coal mining, petroleum exploration and production, timber, or wind/wave farming.

The next set of industrial sectors process energy for society in the form of (1) electrical utilities that produce electricity from burning, coal, uranium, or petroleum, and (2) oil refineries that process petroleum into gasoline, diesel fuel, and petroleum lubricants.

The next set of industrial sectors distribute energy to society through (1) electrical power transmission networks, (2) gasoline and diesel petroleum distribution stations, (3) fuel oil distribution services, and (4) natural gas distribution networks.

Through this complicated scheme of natural cycles and industrial chains, society acquires energy from nature.

LONG WAVES OF ECONOMIC CYCLES

The creation of industrial structures based on new technology has been one of the major factors in modern economic development. This can be seen in the long waves of modern economic history. Modern economies are dynamic systems with cycles of economic expansion and contraction, with both short- and long-term cycles. Many factors contribute to economic dynamics, such as money supply and quality, government regulations, international trade, resource availability and cost, labor supply, cost and skill, market infrastructure, communications and transportation infrastructures, and educational infrastructures. One major factor in the short-term cycles is the managing of product inventory. In contrast, the major factor in the long-term cycles is technological innovation. We will focus on the long-term cycle; the first description of this was by a Russian economist, Leontieff Kondratieff.

Kondratieff's personal story is terrible. He lived in the USSR under the dictator Stalin and perished in the 1930s because of his economic studies on technology and economy. He violated a cardinal rule of communist dictatorships: He told the truth. For this, Stalin had him killed. According to the dogma of Marxism, modern history was supposed to be a struggle between capitalists and labor; when capitalists dominated, they would starve labor, leading to the inevitable decline and collapse of capitalism. Marxists assumed that capitalists would never allow labor sufficient wages to create a market to sustain capitalism over the long term.

Kondratieff decided to test this hypothesis empirically by measuring the economic activity of England. He found, instead of inevitable decline, cycles of expansion and contraction and, overall, increasing economic activity. He asked the question of how capitalism in England was being periodically renewed. Kondratieff's answer was—technology. He plotted a correlation between times of basic innovation and times of economic expansion. Overall, capitalistic economies were expanding rather than contracting because of new technology.

When Kondratieff published his findings in the USSR, Stalin's police arrested him. They shipped him off to Siberia to work as slave labor in the mines of a socialist economy. Kondratieff perished from hunger or disease or brutality—no one really knows—as did 20 million other people that Stalin murdered in his Siberian gulags.

In the 1940s, Kondratieff's work was rediscovered by an American economist, Joseph Schumpeter:

> [T]he work of Schumpeter . . . put emphasis on innovation and on the subsequent burst of entrepreneurial investment activity as the engine in the upswing of the long cycle, à la Kondratieff. This was hesitantly supported by Kuznets, who added . . . the investment burst must stem from innovations with a far-reaching impact across the whole economic system . . .
>
> —(Ray, 1980, pp. 79–80)

The "far-reaching impact" of technology on whole economic systems can occur when new basic technologies create new functionality through new industries or pervade existing industries. In 1975, Gerhard Mensch in Germany again revived Kondratieff's ideas (Mensch, 1979). Following Mensch, Jay Forrester, Alan Graham, and Peter Senge in the United States argued that basic inventions and innovations underlay the beginnings of economic long cycles (Graham and Senge, 1980).

TABLE 2.1   New-Technology Industries and Early Economic Expansion

1770–1800 Economic Expansion

The beginning of the industrial revolution in Europe was based on the new technologies of steam power, coal-fired steel, and textile machinery.

1830–1850 Economic Expansion

The acceleration of the European industrial revolution was based on the technologies of railroads, steamships, telegraph, and coal-produced gas lighting.

1870–1895 Economic Expansion

Contributions to continuing economic growth were made by the new technologies of electrical light and power, the telephone, and chemical dyes and petroleum.

1895–1930 Economic Expansion

The new technologies of automobiles, radio, airplanes, and chemical plastics contributed to further economic growth.

Case Study: Economic Long Waves of the Nineteenth and Twentieth Centuries

In 1990, Robert Ayers brought up-to-date the empirical correlation between European industrial expansion and contraction and the occurrence of new technology-based industries (Ayers, 1990). Table 2.1 summarizes the empirical data.

The invention of the steam-powered engine required the science base of the physics of gases and liquids. This new scientific discipline of physics had provided the knowledge base for Neucommin's and Watt's inventions. We also noted that coal-fired steel required knowledge of chemical elements from the new science base of chemistry which began to be developed in the mid-1700s. Thus, the new disciplines of physics and chemistry were necessary for the technological bases of the industries of the first wave of economic expansion from the beginning of the industrial revolution in England.

The second economic expansion, to which the telegraph contributed, was based on the new electricity and magnetism that were explored and understood in the late 1700s and early 1800s in the new discipline of physics.

In the third long economic expansion, the new physics of electricity and magnetism again provided the phenomenal bases for the inventions of electrical light and power and the telephone. Advances in the new discipline of chemistry provided the knowledge basis for the invention of chemical dyes. Artificial dyes were in great economic demand because of

TABLE 2.2   Economic Cycles in England from 1792 to 1913

The English Economy Expanded from 1792 to 1825, but Contracted from 1825 to 1847.

Kondratieff assigned iron, steam power, and textile machinery as the basis of the expansion, but cited temporary excess capacity as triggering economic contraction.

There Was a Second Period of Economic Expansion in England from 1847 to 1873, Followed by a Contraction from 1873 to 1893.

The expansion was due to the beginnings of new industries in railroads, steamships, telegraph, and coal gas but, again, temporary excess capacity triggered a following economic contraction.

There Was a Third Period of Expansion from 1893 to 1913, Which Was Due to the New Technologies of Chemical Dyes, Electrical Lighting, Telephones, and Automobiles, Followed by Continuing Expansion after World War I, and Then Followed by a Worldwide Depression in the 1930s.

When the First World War began, English economic output turned to war materials. A decade after World War I, England and all Western economies fell into a deep depression from 1930 to the beginnings of World War II. The sources of this depression were complicated—partly temporary excess production capacities and partly excess war debt of the German economy.

---

the expansion of the new textile industry. With these dyes, the gunpowder industry began to expand into the modern chemical industry.

The fourth long economic expansion was fueled by the invention of automobiles, which depended on the invention of the internal combustion engine. This invention required a knowledge base from chemistry and physics. Radio was another new invention based on the advancing science of electricity and magnetism in physics. Chemical plastics evolved from further scientific advances in chemistry.

The major economic expansions in the world were derived from major new science based technologies; but Kondratieff also observed the successive economic contractions in England, as summarized in Table 2.2.

MODEL OF THE ECONOMIC LONG WAVE

Why did economic contractions occur between these economic expansions based on new-technology industries? Why didn't the English or European economies continue to grow smoothly as the new technologies created new industries? The answer that Kondratieff proposed was the following:

Even in a new-technology-based industry, as the industry is ex-panding, production capacity can temporarily outstrip market de-mand, thus inducing excess-capacity contractions.

This kind of long expansion and contraction is what is called a "Kondratieff economic long wave." The stages in this process are as follows:

1. Science discovers phenomena that can provide nature for manipulation by technological invention.
2. New basic technology provides business opportunities for new industries.
3. A new high-tech industry provides rapid market expansion and economic growth.
4. As the new industry continues to improve the technology, products are improved, prices decline, and the market grows toward large volume.
5. Competitors enter the growing large market, investing in more production capacity.
6. As the technology begins to mature, production capacity begins to exceed market demand, triggering price cuts.
7. Excess capacity and lower prices cut margins, increase business failures, and raise unemployment.
8. Turmoil in financial markets may turn a recession into a depression.
9. New science and new basic technologies may provide the basis for a new economic expansion.

History is not deterministic. There is no historical inevitability in Kondratieff's long-wave pattern. It can begin only after scientists discover new nature and after technologists invent new basic technologies. Basic technological innovation can then provide business opportunities for economic expansion; but there is no guarantee that new science and new technologies will always be invented.

What is economically likely is the excess competition and lowered prices as any new technology-based industry matures. So, the important point about the long-wave pattern is that one should expect eventual overproduction, even in a new high-tech industry, as technology matures and many competitors enter the new market. This will always cut profit margins, even in a relatively new industry. High-tech industry will never be high-profit-margin for long—only until technology begins maturing and competition intensifies.

Case Study: Three Hundred Years of Industrialization

Although Kondratieff long waves have been operating throughout the world, the way different nations have industrialized has markedly differed. The historical pattern is this: From about 1765 to about 1865, the principal industrialization occurred in England, France, and Germany. From 1865 to about 1965 (the second hundred years), other European nations began industrializing; but the principal industrialization shifted to North America. By the middle of the twentieth century, in the 1940s, U.S. industrial capacity alone was so large and innovative as to be a determining factor in the conclusion of the Second World War.

For the second half of the twentieth century, U.S. industrial prowess continued, and European nations rebuilt their industrial capabilities that had been destroyed by the war. But for the pattern of industrializations, significant events occurred in Asia. From 1950 to the end of the twentieth century, several Asian countries began emerging as globally-competitive industrial nations: Japan,[1] Taiwan, South Korea, and Singapore. Other Asian countries were also moving toward globally competitive capability: the Philippines, India, China, and Indonesia.

In summary, one can project a pattern of three hundred years of world industrialization, in which different regions of the world began to develop globally competitive industrial industries:

1. First hundred years (1765–1865)—Europe
2. Second hundred years (1865–1965)—North America
3. Third hundred years (1965–2065 [est.])—Asia

The way the different regions industrialized was very different. In the case of the industrialization of Europe, the politics of "colonialism" played an essential role. English trade with nonindustrialized regions was militarily enforced as a colonial empire of the British nation. England accumulated wealth by trading manufactured goods to nonindustrialized nations for raw materials or agriculturally grown materials. The industrial technologies gave England the economic advantage of technologically-based "productivity" in this exchange of goods. Advanced military technologies gave England the power to enforce this trade.

"Colonialism" was, in fact, an age-old way for one nation to exploit another nation by militarily subduing it. All ancient empires were built on military conquest—Egyptian, Hittite, Persian, Macedonian, Roman, Chinese, Mongol, and so on. From 1800 to 1960, European nations were the dominant military colonizers in the world, simply because they had the dominant military power in the world. Accordingly, the first hundred

1. Asian industrialization had actually begun in Japan in 1865, but was diverted principally to a military-dominated society and production; after the Second World War, reindustrialization of Japan occurred. Nevertheless, the major growth of world economies and industrialization was the focus of growth in Asia for the second half of the twentieth century and into the twenty-first.

years of industrialization in Europe, from 1776 to 1876, used colonialism as the export arm of the industrializing European economies.

While European nations continued to industrialize from 1876 to 1976 (and fight major wars among themselves), North America began an industrialization from about 1850 to 1950. U.S. and Canadian industrialization did not use a colonialist model. They had a vast internal landscape, resources, and markets to grow on. North American industrialization used *internal markets* for economic expansion. This lesson shows that industrialization can proceed based on internal markets and does not need the militaristic exploitation of national neighbors.

After 1950, the industrialization efforts in Asia scaled up in scope and quality—first with Japan rebuilding its war-destroyed economy, then with Chinese exiles industrializing Hong Kong and Taiwan, and then with South Korea and Singapore industrializing. Although Japan began modern industrialization in the 1900s, the military culture of its society led its first industrialization to a military economy dedicated to colonial expansion. After 1950, Japan rapidly reindustrialized into a commercially strong economy and strong international economic competitor. In the 1990s, the industrialization of India, the Philippines, Indonesia, and Thailand was proceeding vigorously. In the 1980s and 1990s, mainland China began to industrialize, with capital inputs from the Hong Kong and Taiwan Chinese.

The Asian industrialization from 1950 to 1990 proceeded on a model of importing technology from the United States and Europe and exporting manufactured goods to the United States and Europe. This was encouraged by the United States due to its focus on fighting a cold war against Communist governments from 1948 to 1988.

GLOBALLY COMPETITIVE INDUSTRIALIZATION

In the 1980s and 1990s, there was a growing appreciation of how the continuing industrialization of the world was affecting international business competition. This has begun to be called "globalization"—meaning that the world markets were becoming globally available and production was being distributed globally. Although the entire world is industrializing, it is important to clarify the difference between globally effective and ineffective industrialization. Michael Porter has identified several factors in effective national competitive structures, including political forms, national and industrial infrastructures, domestic markets, and firm strategies (Porter, 1990).

One of the major lessons in the history of the twentieth century is that political dictatorships (of the right or left) were all unable to create industrialized economies that were (1) globally competitive, (2) peacefully sustainable over the long term, or (3) environmentally conservative. A democratic form of government is necessary for long-term economic development. The totalitarian forms of government, Communism and Fascism, were able to provide short-term gains in industrialization, but at a high human price that precluded long-term competitive industrialization. Dictatorships use and promote intellectual dishonesty, economic corruption, and physical brutality. Intellectual integrity and truth are essential for scientific and technological progress. Economic corruption impedes the development of an effective marketplace, substituting graft, shoddy products, and inefficiency for competitive prices, product quality, and productivity. Physical brutality degrades human relationships and creates a culture of terror and cowardliness, which is contrary to the safe environment and courage required for a competitive economy and free society.

An effective national infrastructure is also a necessary condition for effective industrialization. Elements of necessary national infrastructure include educational systems, police and judicial systems, public health and medical systems, energy systems, and transportation and communication systems.

Strategic interaction between universities and high-tech companies is an important feature for industrial competitiveness. For example, Peter Gwynne has described some of the science and technology parks developed in Singapore, South Korea, and Taiwan to build their science and technology infrastructure for high-tech industries (Gwynne, 1993). The model for such science and technology parks was the history of the geographical region of Silicon Valley in northern California in the United States for the building of the computer chip industry and personal computer industry. Two universities, Stanford and the University of California at Berkeley, played an important role, along with venture capital firms, in growing high-tech industries in chips, computers, and multimedia.

Related to national infrastructure but distinctly different is the requirement for an effective industrial infrastructure. An industrial infrastructure requires elements such as financial industry, power industry, transportation industry, communications industry, health industry, and education industry. Each of these industries must have an industrial structure that interacts with and is competitive with international industrial structures.

Successful industrialization has occurred only when the industrialization provided sufficient employment opportunities to create a effective home market for the industrial goods *in addition to* creating successful export products. This has been one of the more puzzling aspects of industrialization—how to create home demand and exports at the same time.

Finally, the strategies of individual firms and the nature of their competition also affect industrialization. For example, the socialization policies of India from the 1950s through the 1980s fostered inefficient and internationally noncompetitive firms by encouraging local monopolies and the purchase of market share by political influence and bribery. As another example, John H. Dunning pointed out the important role for multinational enterprise in building the technological and industrial capacities of developing nations and argued that the motivations of the multinational enterprises was an important factor in whether or not a multinational contributed positively to a nation's development (Dunning, 1994).

In summary, economic industrialization and development pose a complex problem. Technological progress makes it possible for some nations to capitalize and implement economic development. However, which nations succeed in utilizing technology for economic progress has depended on a set of factors, including political system, national and industrial infrastructures, domestic demand, and local firm competitiveness.

Case Study: Whirlpool Becomes a Global Company

In 1989, a U.S.-based firm, Whirlpool, decided to become a global company. This case illustrates the difference in management orientation between a company focused on a national market and a company refocusing on global markets. The historical setting is the late 1980s and early 1990s, when appreciation of the new dynamics of international competition was shifting to a global marketplace as the world continued to industrialize.

The strategic change in Whirlpool's enterprise vision began when David R. Whitwam become CEO of Whirlpool in 1987. Whirlpool was then a North American company, one of the four major competitors in the North American consumer major home appliances market: refrigerators, washing machines and dryers, dishwashers, ranges. In 1987, the industry was a mature technology industry with a saturated market in North America, sales being for replacements and new family startups. It was a substantial market, but with low profit margins and intense competition. As Whitwam recalls:

> Even though we [Whirlpool] had dramatically lowered costs and improved product quality, our profit margins in North America had been declining because everyone in the industry was pursuing the same course and the local market was mature. The four main players, Whirlpool, General Electric, Maytag, Electrolux, were beating one another up every day.
> —(Marcua, 1994, p. 136)

The first thing Whitwam did as new CEO was convene his managers to plan Whirlpool's future:

> When we sat down to plan our future in 1987, it was the first time Whirlpool had ever asked itself what kind of company it wanted to become in the next decade or the next century. This lack of self-scrutiny isn't as surprising as it might sound. Whirlpool was successful, profitable, and reasonably secure in a domestic market that was already eliminating the marginal competitors. The world hadn't broken down our doors the way Japanese auto makers had stormed Detroit, for example.
> —(Marcua, 1994, p. 137)

In a large, successful company, planning usually means to continue to do more of the same thing that first made the company successful. That is reasonable, except when new competitors "storm the castle" or when management cannot project continuing as successfully as it had been in the past. Whitwam saw three choices for Whirlpool's future:

> We could ignore the inevitable—a decision that would have condemned Whirlpool to a slow death. We could wait for globalization to begin and then to try to react, which would have put us in a catch-up mode, technologically and organizationally. Or we could control our own destiny and try to shape the very nature of globalization in our industry.
> —(Marcua, 1994, p. 138)

The first step for Whirlpool to become global was taken in 1989, when Whirlpool purchased N. V. Philips' floundering European appliance business for $1 billion. This acquisition vaulted Whirlpool into the number-one position worldwide in the appliance business. But the acquisition of the Philips appliance business was merely the first challenge to become a global company:

> When we acquired Philips . . . Wall Street analysts expected us to ship 500 people over to Europe, plug them into the plants and distribution systems, and give them six months or a year to turn the business around. They expected us to impose the "superior American way" of operating on the European organization. If you try to gain control of an organization by simply subjugating it to your preconceptions, you can expect to pay for your short-term profits with long-term resistance and resentment.
> —(Marcua, 1994, p. 139)

Whitwam set about to build a global company. Six months after acquiring the European operation, Whirlpool sent 150 senior managers to Montreux, Switzerland, to spend a week developing Whirlpool's global vision. Then, they were given the assignment to educate all of the 38,000 employees of Whirlpool around the world:

> [We] have many, many employees in our manufacturing plants and offices who have been with us 25 or 30 years. They didn't sign up to be part of a global experience. . . . And a lot of our Italian colleagues didn't join Philips to work in the United States. . . . Suddenly we give them new things to think about and new people to work with. We tell people at all levels that the old way of doing business is too cumbersome. Changing a company's approach to doing business is a difficult thing to accomplish. . . .
> —(Marcua, 1994, p. 139)

Whirlpool's decision to globalize followed three principles:

1. Identify common and world-class technologies for the global company;
2. Customize products based on these technologies for local markets; and
3. Organize on an interactive, global basis.

A product uses common technology, but products needed to be customized to the regional market:

> Washing technology is washing technology. But our German products are feature-rich and thus considered higher-end. The products that come out of our Italian plants run at lower RPMs and are less costly.
>
> —(Marcua, 1994, p. 136)

To organize on a global basis, Whirlpool developed new procedures for the company based on international participation. For example, one of the assignments to managers at Montreux was to commission 15 projects on management and technology that each company in Whirlpool was to develop for the whole of Whirlpool, which they called "One-Company Challenges." Common procedures for the new global Whirlpool was a first step in globalizing the company.

For example, one of the "One-Company Challenges" was to create a company wide total quality management system. When Whirlpool acquired Philips, Philips was operating on the ISO 9000 approach to quality, generally adopted in Europe. However, U.S. Whirlpool had been using the U.S. Baldrige approach to quality. Instead of imposing one system on all, a cross-cultural team was given the task of examining the best of both approaches and forging a new global quality system for Whirlpool.

The vision of Whirlpool becoming a global company was based on a desire to access all the markets of the world and to create and maintain a dominant position in each local market.

GLOBAL COMPETITIVENESS

Any large company trying to survive competitively in the coming twenty-first century should manage with both global and local considerations. The famous slogan for this philosophy came from Sony management, which urged, "Think globally and act locally."

This meant that to be competitive in the global market requires:

- World-class technology
- World-class finance

- World-class production
- World-class communications and transportation

At the same time, consideration should be given to:

- Products that are at the same time both world-class and locally focused
- Local markets
- Local distribution

Production can also be localized to employ local people and local management. This is important, because local employment provides the basis for consumer incomes to purchase the products of a company. Integrating local management into an integrated global firm is a principal challenge. For example, Alan David MacCormack, Lawrence James Newman III, and Donald B. Rosenfield examined international trends in strategy for locating manufacturing sites in a global market and suggested that the best sites will be decentralized manufacturing plants in large, sophisticated regional markets (MacCormack et al., 1994).

Hideo Sugiura described Honda Motor Company's four principles for localization in a global economy: products, profits, production, and management (Sugiura, 1990). Localization of products means developing products suited to the actual and potential needs of customers in a specific region. Localization of profits means reinvesting some profits in the local market to regard each regional unit as a kind of local company. Localization of production means providing employment in the regional markets to which a global company sells. Localization of management means developing local employees and management that share corporate philosophy but work locally.

TECHNOLOGY INTELLIGENCE

To gain a competitive advantage from technology, one must not only acquire the best technology in the world, but also improve upon it. This means that "technology transfer" from one country or firm to another is insufficient to become globally competitive. When a technology is transferred, it is usually what is competitive today but not what will be competitive tomorrow. To provide a competitive difference, technological information must be both timely and utilized.

For example, Atsuro Kokubo emphasized the need not only to look around for technology, but also to integrate that information search with innovativeness:

> Japan's large corporations have a well-deserved reputation for aggressiveness in gathering competitive technology information worldwide. . . . In many industries, the Japanese are steadily changing from imitation to innovation. . . .

In keeping with this switch, the intense global R&D competition that is emerging will require companies to adopt a more integrated style of information gathering. The search for technology information must be more closely joined to competitive intelligence activities for R&D.

—(Kokubo, 1992, p. 33)

Competitive intelligence for R&D requires that technical information be gathered on a worldwide basis, be distributed internally in a company, and used for technology strategy. Kokubo also emphasized the integration of intelligence into technology strategy:

Competitive intelligence activities must enable a company to evaluate its own technological competitiveness. . . . This analytical drive spreads a dramatically new R&D philosophy throughout a company.

—(Kokubo, 1992, p. 34)

One important result of this globalization is that there is now a trend for global companies to disperse their R&D efforts throughout the world. For example, Manuel G. Serapio studied the growth of R&D investments by U.S. and Japanese firms outside their home countries. The reasons cited for this globalization of R&D were to improve product focus on regional markets, monitor technology developments, and acquire new technology (Serapio, 1995). In another example, Vittorio Chiesa examined the international dispersion of R&D laboratories of 12 companies in the early 1990s and found the global R&D dispersion was increased for the labs taking long-term horizons (Chiesa, 1996).

In the 1990s, the trend toward internationalization of a firm's R&D was relatively new and still small. For example, Pari Patel examined the patenting records of 539 firms in several industries and found that the overwhelming majority of technology invention still occurred in laboratories in the company's home country (Patel, 1996). But this was the trend—dispersion of the capability of innovating new technology around the world—not only a global market, but also a global R&D system.

TECHNOLOGIES OF POWER AND OF THINKING

Another way of looking at the structuring of industry from the continuing industrialization of the world is to examine the general types of technologies developed in the nineteenth and twentieth centuries. Broadly speaking, one can call most of the technologies developed in the nineteenth century as "technologies of power," in contrast to the twentieth century, when a new broad category of technologies began, which one could call "technologies of thinking."

We recall that the early driving technologies of the industrial revolution in England were the inventions of the steam engine, coke-based steel production, and textile machinery. One can call these technologies of power in the sense that energy was being used to create and process materials. In addition, in the early 1800s, the steam engine was applied to railroads and steamships to provide new technologies of transportation. We can call such technologies of energy, materials, processing, and transportation technologies of power.

While these technologies continued to develop in the twentieth century (such as nuclear power, plastics, molding, and flight), there also arose from the middle of this century a new group of technologies around the computer, which we could call the technologies of "thinking." These were communications, information, computation, and decision technologies. For example, the technologies of the telephone (begun in the nineteenth century), radio, satellite, and fiber optics (all twentieth century) enabled new functionalities in communication. The invention of the computer during the Second World War provided new functionalities in storing and manipulating information, doing computations, and enabling decision and control capabilities.

What is important about these technologies of power and thinking is to recognize that not only has the industrialization of the world been continuing over the last two centuries, but also new categories of technologies have been added. This means that how nations compete economically in the twenty-first century will be different from how they competed in the nineteenth and twentieth centuries. Thus, "globalization" can mean not only *where* we compete in the twenty-first century, but *how* we compete. Competition through communications, information, computation, and decision making is now as important as competition through energy, materials, processes, and transportation.

SUMMARY

An innovation of a new basic technology creates a new industrial structure; the industrial structure is organized around the sequence of technologies needed to create the value-addedness of the new functionality that connects nature to consumer. These industrial structures have been a major factor in long-term economic development, resulting in waves of economic expansion and contraction called "Kondratieff long waves." These waves have played an important role in the industrialization of the world.

The world's industrialization has been uneven, occurring at different times in different regions of the world. As the twenty-first century begins, competitive industrializations are progressing rapidly in Asia. Unfortunately, in the past, not all national attempts to industrialize have resulted in internationally competitive industries. Several factors that are important in successful international economic competitiveness include the political system, national and industrial infrastructures, domestic market demand, and firm rivalry.

As the world eventually industrializes, the nature of international economic competition has been changing; this has been called "globalization." Principles for successfully globalizing a firm include world-class technologies for the global company, regionally customized products, and organization on an interactive international basis. In addition, new technologies of power and of thinking mean that how nations compete economically in the twenty-first century will be different from how they competed earlier.

FOR REFLECTION

Choose a modern nation and describe its transformation into a industrialized economy from 1800 to 1990. What have been the political, economic, and social problems in this transformation? Identify the kinds of impacts that technologies made on these.

____3
TECHNOLOGY

CENTRAL CONCEPTS

1. Systems concept
2. Technology system
3. Loci of change in a technology system
4. Application system
5. Role of technology in altering society
6. Scales of innovation change
7. Discontinuous and continuous technological change
8. Multiple meanings of the term 'technology'
9. Patents

CASE STUDIES

Sikorsky's Helicopter
Monster Guns of the Nineteenth Century
Innovations in Railroads

INTRODUCTION

Every technology is a system. It is this systems concept of technology that allows one to plan technological progress. Through technology planning, one can create and support research programs to advance technology. Accordingly, the key to successful technology innovation begins with conceiving of technologies as systems.

The basic management problems about technological innovation are (1) how to foster important inventions and (2) how to develop these into commercially successful innovations. These problems are not easy to solve, as the cold facts about technological innovation tell. Very few inventions are novel and important. Of these, very few inventions ever get innovated. And of these, very few innovations are commercially successful.

In this chapter we will address the following questions:

- What is the general concept of technology as a system?
- What are the general elements of a technology system that can be used to manage technical progress?
- How can one technology give rise to several technology systems and different applications?
- What determines the scale of economic impact of a technical change?

TECHNOLOGY FUNCTION AND APPLICATION

The purpose to which a technology is put is called its "application"; the ability to do something for that application is called its "functional capability"; and how well the technology performs with that ability for the application is called its "performance." Until the performance of a new technology is sufficient for a customer's application, the technology is not yet ripe. The critical performance measure of a technology is the minimum performance necessary for the technology to do an application for a customer. Accordingly, the aspects of a new technology that management should begin to think about are function, customer, application, and critical performance.

"SYSTEM" CONCEPT

The concept of a "system" means to look at a thing, an object, with a view to seeing it as a "totality," displaying "change," and encompassed in an "environment."

A description of a thing as a system captures its "totality" as transformations of system states within a relevant environment.

There are two general forms of systems, closed and open systems (Betz and Mitroff, 1974). A closed system does not have significant outputs from its environment and may also not have significant inputs. A closed system is principally described by transformations between internal states of the system. For example, the earth as an isolated planet sustaining life can be viewed as a closed system, wherein the only significant input is energy from the sun and is describable as closed cycles of physical and living processes (except when occasionally impacted by asteroids).

An open system has both inputs from its environment and significant outputs into its environment. Both businesses and technologies are open systems. A business, viewed as a system, is a legal and financial entity, purchasing resources from its economic environment and selling products and/or services into its economic environment. A technology, viewed as a system, is a functional totality, transforming inputs from its purposive environment into means-outputs of its purposive environment.

Case Study: Sikorsky's Helicopter

This case study recounts the invention of the rotating-wing form of powered flight, the helicopter. Powered flight was not possible until the invention and innovation of the gasoline-fueled internal combustion engine. Four hundred years earlier, the notebook of the artist, Leonardo da Vinci, contained sketches for a rotating-wing kind of helicopter, showing that others had envisioned such flight. But the difference between science fiction and technology is that inventions must be real and able to work.

The historical setting of this case is the early twentieth century, soon after the Wright brothers demonstrated the first powered fixed-wing flight in 1904 at Kitty Hawk, North Carolina. In the following years, they demonstrated their invention around America and in Europe. Then, many other inventors and developers began improving the airplane and inventing other schemes; one was Igor Ivanovich Sikorsky.

Sikorsky was born in Kiev, Russia, in 1889. Even as a child, he showed the technical bent of an inventor (Wohleber, 1993). At the age of 14, Sikorsky entered the Imperial Russian Naval Academy in St. Petersburg. He realized he did not wish to become a naval officer, and quit after three years to enroll in the Kiev Polytechnic Institute. In the summer of 1908, he visited Germany and learned of Count von Zeppelin's dirigibles and of the powered flights of Orville and Wilbur Wright, who were then touring Europe with their amazing demonstration of powered flight. Sikorsky determined that his life's work was to be flight. Flight was truly a new functional capability for humanity.

From the beginning, Sikorsky wanted to do something different from the Wright brothers' kind of plane. He resolved to design a flying machine that would be capable of "rising directly from the ground by the action of a lifting propeller." His idea came from childhood remembering of his mother's descriptions of Leonardo da Vinci's designs for flying machines. In a German hotel room in 1908, he put a four-foot propeller on a small engine and found it could achieve a lifting force of 80 pounds for each horsepower the engine could provide.

Back in Kiev, he built two helicopter prototypes, but the French-built, three-cylinder 25-horsepower engine he had purchased did not provide sufficient power to lift the helicopter. He then turned to constructing fixed-wing aircraft. With his fifth prototype, he got one to fly. After this, Sikorsky built a sixth prototype, a three-seat bi-wing plane that flew at 70 miles an hour, breaking the then-world record for speed of a plane carrying a pilot and two passengers. He built the world's first four-engine aircraft; in May 1913, he flew the four-and-a-half-ton plane above St. Petersburg. He then built a refined, larger version of this; and when World War I began, he was put in charge of building military versions for long-range reconnaissance and bombing raids. By 1917, he had produced seventy planes of his "Il'ya Muiromet."

Czar Nicholas of Russia abdicated his throne in 1917. Soon, a small group of Communists seized power in Russia. Many Russians were forced to flee the new dictatorship, and Sikorsky left in February 1918 (leaving behind the money he had made). He traveled to France and presented plans to build bombers. The French government accepted them, but the First World War ended before Sikorsky could begin production.

In March 1919, Sikorsky traveled to the United States. For the next few years, Sikorsky barely earned a living as a teacher. In 1923, he started a new airplane firm and built an all-metal passenger plane with two 300-horsepower engines. (It was eventually sold to a filmmaker, and, disguised as a German bomber, went down in flames in a Hollywood film.) In 1928, Sikorsky built a large amphibious plane, the S-38, which sold well. The largest customer was Pan American Airways, which flew 38 of the S-38s on its Latin American routes. Pan American called them the "American Clipper." Sikorsky then built larger versions of the plane; the S-40 was the first to cross the Pacific. Sikorsky eventually sold his company to United Aircraft.

In 1938, Sikorsky went back to his first technical love and began working on a design for a helicopter for United Aircraft. He patented a design that used a larger horizontal propeller for lift and a small vertical one on a tail to control torque. The technical problem then was to provide control. (The fixed-wing aircraft, properly designed, provides relatively stable flight dynamics, compared to the helicopter.) It wasn't until 1940 that Sikorsky could develop a prototype that could hover steadily. Then the Army got interested in the project.

On May 6, 1941, Sikorsky demonstrated a new prototype, the VS-300, to reporters and military representatives. Sikorsky took the helicopter to a stationary hover and remained there for one hour and 32 minutes. The helicopter still flew forward too slowly, yet it was enough for a military R&D contract. The Army wanted a higher forward speed and insisted that all control be in the main rotor and that there be an enclosed cockpit with a passenger seat.

By 1942, Sikorsky demonstrated a military version, the XR-4, which had a 165-horsepower engine and could both hover and travel forward at a speed of 75 mph. Sikorsky then began producing helicopters for the Army as models R-4, R-5, and R-6. By the end of World War II, four hundred military helicopters had been produced. (In March 1944, an R-4 was used for the first military rescue mission, lifting an American pilot and three wounded British soldiers from a crash site in a rice paddy in Burma. Later, during the Korean war, helicopters would be extensively used for rescue. And later, during the Vietnam war, helicopters would also be used to deploy ground troops under combat conditions.)

For the innovation of the helicopter, the following conditions were required: (1) the prior invention of the gasoline engine, (2) the prior invention of fixed-wing powered flight, (3) the development of engines that were sufficiently light and powerful to power a helicopter, (4) the dedication of an inventor to solve the practical problems of controlled hovering and flight, and (5) the sponsorship of a military customer to fund the R&D required to perfect a practical prototype for application.

Sikorsky retired from Sikorsky Aircraft in 1957 at the age of 68. But he remained a consultant for the continuing development of helicopters until he died in 1972.

NATURAL PHENOMENA AND A LOGIC OF MANIPULATION

We recall that the definition of technology is a knowledge of manipulation of nature for human purposes. If one examines any technology, one will find these ingredients:

1. Natural phenomena
2. Manipulation of natural states
3. A logic of human purpose

All technologies are based on natural phenomena. For example, in flight, it is the natural phenomenon of the physics of airflow over an over-head propeller (helicopter) or over a horizontal wing (fixed wing) that provides lift. The physics of helicopter lift is simply action-reaction, whereas the physics of airplane wing lift is the differences in air pressure as the air flows over the curved upper part of the wing and under the flat lower part of the wing. Different ways of manipulating natural states can produce different versions of a technology (e.g., helicopter and airplane versions of flight technology).

In addition to being based on nature, all technologies are also based on human purpose. The reason one manipulates nature in a technology is to achieve a purpose, a human goal. For example, the purpose of flight technology is for transportation through the air. In technology, human purpose is expressed through the *logic of the manipulation.*

The human purpose of the manipulation of a technology is embedded in the logical schematic of the technology.

Technologies are, therefore, mappings of a schematic logic that expresses the functional transformation against a sequence of phenomenal states that provides the natural basis of the technology. These two aspects can be called the "schema" and "morphology" of a technology system. The inventive creation of a new technology is a devising of (1) a schema, (2) a morphology, and (3) a one-to-one mapping between schema and morphology.

TECHNOLOGY SYSTEM

A technology system is a configuration of parts whose operation together provides a *functional* transformation. Any technology might be invented in different configurations. Any particular technology system is a *specific configuration* of a technology focused by an application. Different technology systems provide the similar functional capability but with different features and performance. For example, in the function of flight, there were two different configurations to attain flight, airplanes and helicopters; these are called different technology systems (different configurations of a technological function): fixed-wing and rotating-wing.

LOCI OF CHANGE IN A TECHNOLOGY SYSTEM

Since any technology system has a specific schematic logic and morphological architechture, change can occur in logic or morphology.
    Change can occur in the schematic logic of the technology system or in the strategy of its applications system. For example, the innovations of airplanes and submarines both dramatically altered naval military strategies.
    The general morphological form of any technology system consists of:

1. Boundary of the system
2. Construction material of the system
3. Parts of the system
4. Connections between parts of the system
5. Control subsystem

Change can come from altering the boundary of a technology system. For example, in the 1980s, computer system performance was improved by expanding the boundary of the computational system from a single computer to a network of connected computers.

Change can occur in the construction material of a technology system. For example, in the 1970s, airplane performance was improved by substituting composite materials for aluminum in construction.

Change can occur in parts of a technology system. For example, in the 1980s, performance of personal computers was improved by generations of parts: the central processing unit (e.g., the Intel 8088, 80286, 80386, 80486, and Pentium CPUs) and the dynamic memory (e.g., memory chips of capacities 8K, 16K, 64K, 240K, 1M, 4M, 16M, etc.).

Change can occur in connections between parts of a technology system. For example, in the 1970s, performance of telephone networks was improved by substituting fiber-optic cables for copper wire cables.

Change can occur in the control subsystem of a technology. For example, in the 1970s, automobile fuel performance was improved by replacing carburetors with electronic fuel injection systems.

Case Study: Monster Guns of the Nineteenth Century

Societal conditions influence when and how an invention occurs. This case study illustrates how the change in the performance of one technology in an application may stimulate a need for change in another technology of the application.

The historical setting is the time of important technological change in the history of military technology—the innovation of the ironclad warship. Control over the oceans had been a critical issue among nations. Although ships could have been clad with iron earlier, they would have been clumsy to sail. The invention of the steam engine had made this practical.

Ironclad ships created a military crisis for the supremacy of Great Britain's navy, which earlier defeated Napoleon's French navy (Bastable, 1992). In 1860, a new Bonaparte was on the French throne, Napoleon II; he had steam engines installed in France's fleet, launching its first ironclad frigate (a mid-sized warship), named "La Gloire."

The British admiralty made a mock-up target of La Gloire's sides—wood clad with four-inch iron plate. They fired on the target with existing naval cannons, but their balls only bounced off the mock-up target. Cannons then were made of cast iron, smooth bore, loaded at the mouth, and fired round iron balls. The military point was obvious: No ship in the British fleet could sink La Gloire. If the French chose to invade England's island with a fleet of ironclads, the British fleet could do nothing to stop them. The prime minister, Lord Palmerston, appointed a Royal Commission on the Defense of the United Kingdom to design a series of coastal fortifications with guns large enough to sink ironclads—"monster guns." A British inventor, Sir William Armstrong, would supply the monster guns for the fort.

Armstrong had pioneered the development of large guns made of layers of wrought iron, as opposed to guns made of cast iron. He added rifling to the layered iron gun, and breechlocks with which to load the gun from the rear rather than down the muzzle. Some of Armstrong's ideas were not original, but borrowed from recent technical advances in shoulder-held guns (which we now call "rifles"). The technical problem Armstrong had to solve was how to make the rifled barrels stronger for cannon and how to devise strong breechlocks for cannon.

His invention was to use a coiled arrangement of strips of wrought iron wound into a barrel and welded into a solid tube. The wound strips provided rifling for the cannon, and layers of wound strips built up around the smallest core provided strength for the explosions in the large cannon to expel the shell. Armstrong's revolutionary large guns had already demonstrated military superiority over existing cast-iron, smooth-bore cannon. He produced these for the British government, and such guns, firing 40-pound conical shells, had demonstrated their long-range accuracy and deadly efficiency in the British participation in the Anglo-French invasion of China in 1860.

But a 40-pound shell could not penetrate four inches of iron plate at 1200 yards, which is what would be required to sink La Gloire. Armstrong calculated that it would take at least a 300-pound explosive shell to penetrate four-inch plate iron from that distance.

In 1862, Armstrong completed his first 300–pounder and demonstrated it at the government firing range at Shoeburyness, at the mouth of the Thames. The next day, the London *Times* published an enthusiastic account:

> With an indescribable crash that mingled fearfully with the report of the gun, the shot struck upon a comparatively uninjured plate, shattering the iron mass before it into little crumbs of metal, splintering the teak (behind that) into fibers literally as small as pins.
>
> —(Bastable, 1992, p. 232)

Armstrong not only supplied the British military, but also sold monster guns to the world market. These provided a major turning point in military technology. Soon, wrought iron was replaced by steel. The German armament maker Krupp perfected the steel cannon, which played an important role in the German defeat of the French in 1875 in the Franco-Prussian war. In 1914, the world began to be subjected to the terrible destruction this new artillery was to produce in the First World War.

The deliberate cooperation between an inventor, such as Armstrong, and government opened a new era in the management of military technology. As the historian Bastable commented:

> Armstrong's first breechloaders and his monster guns set in motion technological, political, bureaucratic, and industrial forces that established a more intimate link between industry and government. This new relationship reflected the fact that the ability to wage war and build empires had come to depend more than ever before on the industrial capacity of an economy and continual technological innovation by its engineers.
> —(Bastable, 1992, p. 246)

APPLICATION SYSTEMS AND ARTIFACTS

Technologies occur in different system configurations; which particular configuration is selected depends on how the technology system is used—its application. An application is also a system, transforming inputs into desired outputs that accomplish a purpose. An application system consists of:

1. A major device system and all the technologies embodied in the device (e.g., the battleship for naval warfare);
2. Key peripheral systems and all the technologies embodied in the peripherals (e.g., naval cannon); and
3. Strategies, tactics, and control technologies for using the major device system and peripheral systems in the application.

The concept of a "technology system" focuses on the techniques for attaining a functional capability. An "application system" focuses on how a functional capability is used by a customer. The concept of a "major-device system" of an application involves the primary technological skill used in an application. The focus of "key peripheral systems" is assisting technologies for the major device system in an application. "Strategies, tactics, and control technologies for an application" focus on how a customer uses a major device system and its peripherals in an application.

Sometimes, the term "technology" is used to cover several of these meanings. Some use the term "technology" to indicate the "inventive" idea of a functional capability. Others use "technology" to indicate the idea of the "system" of the capability—the technology system. And some use the term "technology" to indicate the "context" of the technical application—the application system. Others use "technology" to indicate key "artifacts" used in the application—product systems. However, let us distinguish these different uses of the term "technology":

1. Technology as **invention**—the *generic concept* of technology
2. Technology as **system**—the *different configurations* of technology
3. Technology as **application**—the *different applications* contexts of technology
4. Technology as **artifact**—the *different devices* embodying technology systems that are used in an application

For example, "flight" is a generic concept of a technological capability of transport through air. Flight as a generic technology depends on applying power to provide lift in air. Airplanes and helicopters are different technology systems for flight. Applications of military or commercial transportation systems use airplanes and helicopters as artifacts (major device systems) within the transportation system.

ROLE OF TECHNOLOGY IN ALTERING SOCIETY

Technologies can alter societal structures when they affect a society's:

1. Power structure;
2. Means of production;
3. Organization;
4. Distribution of wealth; or
5. Means of communications.

For example, the gun affected the power structure of societies by changing the demand for skills in using weapons. In the power structure of the dominant warrior caste of the aristocracy, barons and kings had almost equal capability to field feudal armies equipped with swords, bows and arrows, lances, and siege weapons. Swords, bows and arrows, and lances require a great deal of skill to use successfully. Training in these skills takes years. A skilled feudal warrior could defeat several relatively less-skilled warriors. In contrast, the skills to use guns can be attained in a few weeks, but a skilled gunman cannot take out several semiskilled gunmen. Using guns, raw firepower counts over marksmanship. Thus, peasants could be trained into an effective army when given guns instead of swords and bows and arrows. A monarch's ability to raise money with taxes and then pay and equip a gun-armed peasant army became the base of new power to end the feudal age in Europe. In the 1500s and 1600s, monarchies began to systematically reduce their baronies into a dependent service aristocracy without independent military capability.

The steam-powered textile machinery innovation provides an example of how technology can alter both production means and organization in societies. Before the invention of powered textile machinery, thread was spun and cloth woven by hand on spinning wheels and looms in the individual cottages of spinners and weavers. A jobber would periodically bring them materials to be worked and pick up the worked pieces, paying a piece rate. Jobbers played the role of contractors and cottagers the role of individual producers. Both the roles and the organization of production were altered with the introduction of the factory system for the new means of production by powered textile machinery.

The cost of the new machinery necessitated putting the machinery in a central factory and hiring people to come to the factory and attend the machinery. The workers were then paid for their time. Since these inventions were used in a factory organization, factory-based industries changed societal organization from the guild form of feudal organization of production into modern capitalistic forms of production. People left farmland and moved into cities to find employment in the new factories. Capitalist owners and managers became "bosses" and peasants "labor." Societal structure altered from aristocracy and peasantry to management and labor.

The railroads affected the communications and transportation of people and goods. Before railroads, the time and cost of transporting people and goods across land inhibited the development of national economic infrastructures and the growth of national-scale markets. For example, the European colonization of the North American continent rapidly expanded beyond the east coast into the west when the railroad promoted economic interchanges between the manufacturing capability in the east and the agricultural capabilities of the American midwest.

SCALES OF TECHNOLOGICAL INNOVATION

When technology changes, how big an impact does it make on applications? This idea has been called the *scale* of the innovation. Early in the studies of technological innovation, Donald Marquis distinguished different scales of innovation impact: radical, incremental, and systems (Marquis, 1969). However, what Marquis called a radical and a systems innovation has not proven to be a useful distinction, since all technologies are systems. A more useful distinction is that of a next-generation technology (Betz, 1993). Accordingly, we will distinguish three scales of technological innovation:

1. *Radical innovation*—a basic technological innovation that establishes a new functionality (e.g. steam engine or steamboat)
2. *Incremental innovation*—a change in an existing technology system that does not alter functionality but incrementally improves performance, features, safety, or quality or lowers cost (e.g. governor on a steam engine)
3. *Next-generation technology innovation*—a change in an existing tech-

nology system that does not alter functionality but dramatically improves performance, features, safety, or quality or lowers cost, *and* opens up new applications (e.g. substitution of jet propulsion for propellers on airplanes)

Basic inventions occur infrequently in the history of technological innovation, for most innovations are improvements on existing technologies. Although infrequent, basic innovations are extremely important because they create new industries. Basic inventions can be of two kinds, component inventions and system inventions. A component invention is used within a system, and a system invention invents new ways of incorporating components into a system. However, components are also systems. In this way, there occurs a hierarchy of technology systems, containing other technology systems as components.

All basic innovations—component or system—require improvement through incremental innovations. Over time, most innovations are incremental and add up to major improvements in the basic innovation.

Occasionally, a technology system is improved not just in part, but in whole. When this happens, the new whole replaces the older system, and this is called a next generation of the technology (NGT). Next generations of technology systems provide dramatic improvement in performance and features. Incremental innovations provide small but definite improvements in performance or features.

However, the perceived impact of an innovation can depend on the perspective of the user. For example, Allan N. Afuah and Nik Bahram pointed out that what may be an incremental innovation to a supplier may make a larger impact on the supplier's customer (Afuah and Bahram, 1995).

Case Study: Innovations in Railroads

This case illustrates the range of scale of innovations in a technology. The historical context of this case study is the long period of technology evolution of the 1800s through the 1900s in railroads—from their basic invention, using the steam engine, to the eventual technical obsolescence of the steam engine.

The key invention that began the industrial revolution in the eighteenth century was the steam engine. The steam engine was a radical invention that provided the first source of mechanical power.

As a component, the steam engine found several initial applications—pumping water from mines, powering textile machinery, powering railroads, powering steamships—and each application also became a new radical innovation.

The steam engine had many technical improvements, one of which was a two-stroke steam-cycle engine that doubled an engine's horsepower. Early steam engines on railroads had steam pushing the piston on only one side; later, improved models had steam pushing the piston alternately on both sides. This was an incremental innovation improvement that advanced the performance of the railroad.

Steam-powered locomotives dominated into the early twentieth century as a technology, until they were replaced by diesel-engine locomotives. The diesel-engine locomotive is an example of a next generation of a technology system.

IMPACT OF THE SCALE OF TECHNOLOGICAL CHANGE ON INDUSTRY

Both radical and next-generation technological innovation can have dramatic impacts on industrial structure. For this reason, it has been useful to talk about these innovations as kinds of "discontinuities" in technological progress.

Discontinuous technological change creates industrial structures and may alter existing industrial structures.

This is in contrast to the kind of smaller change provided by incremental technological innovations, which tend not to alter industrial structure. For this reason, incremental technological innovation is also called "continuous" technological progress.

Continuous technological change reinforces an existing industrial structure.

INTELLECTUAL PROPERTY

Patent law is very important to technological innovation. The innovation of a new technology is facilitated by allowing an innovator to capture economic benefits of the technological innovation. Patents are important to assure the innovator that the development costs of a new invention can be regained through a temporary legal monopoly to use the invention.

Technical knowledge that is valuable to a firm may be nonproprietary or proprietary. Nonproprietary technology is technical knowledge that is shared widely in society. Generally, scientific and generic engineering knowledge are nonproprietary.

Proprietary technology is technical knowledge that no one but the firm has or may legally use for commercial applications. Proprietary technical knowledge includes trade secrets, copyrights, and patents. These may be legally protected within the area of law called "intellectual property" (which also includes trademarks).

A trade secret is any formula, pattern, device, or information, used in business to provide a competitive advantage, which is held secretly in a business. By holding the information in secret, one needs to establish that it could be obtained by another firm only through some effort that can be seen as invading the privacy of the firm. A copyright is a registered exclusive right to authors and composers or artists, providing control of the commercial use of their forms of expression. A patent is a legally protected right to commercially use a novel and useful invention. A trademark is a recognizable symbol that a firm may use to establish its identity.

Patents are important in exploiting new technology commercially. Copyrights are important in exploiting new software commercially.

In the United States, the most common form of patent is called a "utility patent," which is granted by the U.S. government to inventors. Utility patents are legal rights to exclude anyone other than the inventor (or those whom the inventor has licensed) from making, selling, or using products incorporating the patented idea. U.S. patents are for a specific term, normally 17 years (except for medical products that must pass federal Food and Drug Administration standards). The idea of giving an inventor a legal monopoly on an invention for a specific term originated in Europe as a way for the inventor to disclose full knowledge of the invention to society (Bell, 1984).

To gain a patent, an inventor must file a disclosure with a patent office. To be patentable in the United States, an invention must meet standards of novelty and utility. Filing of the patent includes the conception of the idea and reducing the idea to workable form. Workable form may be an actual device, built and working, or may be full disclosure such that a working model can be built. Generally, information about the invention must not be disclosed to the public before filing a patent application, or the application will be invalidated.

Each country has its own patent laws and offices. To gain legal protection, an inventor must file separately in each country in which the idea is to be protected. Accordingly, filing patents for worldwide protection is very expensive. In the United States, only the inventor may file a patent application. In many other countries, the first to file (whether or not an inventor) may acquire a patent.

In the United States, general organizational practice has evolved to require all employees engaged in research supported by an organization to assign rights to any inventions so made to the organization. This ownership of invention rights by employers has been upheld in U.S. courts, even in the absence of formal agreements between employers and employees about patent rights.

Licenses on patents are rights that a patent owner may grant to others to use an invention commercially, freely, or in exchange for consideration, such as money or property. Licenses may be granted exclusively to one licensee, or nonexclusively to several licensees.

Generally, it is the responsibility of the patent holder in a country to litigate to defend against patent infringement. Defending patents can be a costly and lengthy process.

Leonard Berkowitz emphasized that to obtain commercial value from a patent portfolio, a company should have a patent strategy that integrates with its technology strategy (Berkowitz, 1993). Patent strategies include grouping patents for a defensive line around a technology, obtaining of international patents for global markets, and using patent searches as intelligence inputs about competitor's technology strategies.

Patent policy can make a difference in the directions of commercialization in a new technology. For example, Sandra M. Thomas, Keiko Kimura, and Julian F. Burke studied cases of patent litigation among biotechnology firms and found that the upholding of broad claims led to a curtailment of generation of modified variants covered by the claim (Thomas, Kimura, and Burke, 1995).

SUMMARY

All technologies are systems; this systems aspect makes planning technological progress possible. As a system, any technology rests on a natural phenomenon and on a schematic logic for manipulating the phenomenon. A specific system of a technology is a particular configuration of a phenomenal morphology for a schematic logic of the technology. The loci of change in any technology system can occur in its schematic logic, or its morphological boundary, construction material, parts, connections, or control subsystem. An application system describes how a technology system is used by a customer.

When a new technology system is innovated, the scale of impact on applications can range from radical to incremental or as next-generation technology. Discontinuous technological innovation creates and alters industrial structures; continuous technological innovation reinforces existing industrial structures. Patents are important in commercially exploiting new technology.

FOR REFLECTION

Choose a basic invention that created new functional capability for society. Who invented it, when, and under what societal conditions? Describe the scientific phenomena on which it is based. Describe the technology as a system, including its boundary, construction material, parts, connections, and control subsystem. Sketch the history of its technical progress as a system.

____4
INDUSTRIAL DYNAMICS

CENTRAL CONCEPTS

1. Industrial core technologies
2. Strategic technologies matrix
3. Core technology industrial life cycle
4. Industrial design standards
5. Technological maturity in an industry
6. Industrial competition in technology life cycles
7. Commodity-type products
8. Product-line lifetimes
9. Technical obsolescence in an industry
10. Scientific progress for industrial revitalization

CASE STUDIES

Core Technologies at Nippon Steel Corporation
Development of the U.S. Auto Industry
Product and Process Innovations in the U.S. Auto Industry
Numbers of U.S. Auto Firms over Time
Commodity-Type Products of the Chemical Industry in the 1980s
Product-Line Lifetimes in Tire Chords
High-Tech in the Pharmaceutical Industry

INTRODUCTION

We have seen how, at the macro-economic level, new basic technologies have facilitated the economic development of the world. Next, we will review, at the micro-economic level, how the dynamic processes of industrial development occur. We will address the following kinds of issues:

- What are the impacts of technological change on industrial dynamics?
- What pattern does industrial growth follow during and after technological progress?
- What are the competitive conditions for an industrial structure as technology changes?

Case Study: Core Technologies at Nippon Steel Corporation

This case study illustrates how research organization focuses first on the core technologies of a business. It describes the research organization at a major industrial steel producer, Nippon Steel Corporation. The historical setting is the 1990s, when Nippon Steel was one of the world's largest steel producers.

In 1990, Nippon Steel produced 29 million metric tons of crude steel. It had 54,000 employees and nine steel works across Japan. Two managers of Nippon Steel, Tatsuya Kimura and Makoto Tezuka, then summarized the vision of the company:

> Steel is one of modern civilization's essential basic materials, and Nippon Steel Corporation considers its role to be one of supplying society with quality steels at reasonable prices. To realize this, we carry out state-of-the-art R&D activities regarding steel materials and their manufacturing technologies.
>
> —(Kimura and Tezuka, 1992, p. 21)

For strategic planning, Nippon Steel had a Technical Administration Bureau, which performed planning for steel-related business activities. Nippon Steel had several central research laboratories: The Steel Research Laboratory concentrated on the improvement and development of steel products. The Process Technology Research Laboratory focused on improving and developing iron- and steel-making processes. About 1200 people performed research and development work, and about 7000 were engaged in the development of steel materials production technology at the steel works, and other technical fields.

Yet, despite the size and importance of the steel materials business, in 1992, Nippon Steel was planning business diversification into other new businesses and technologies:

Nippon Steel's plans are to achieve diversification so that in the near future steel will account for less than 50 percent of the company's total sales.

—(Kimura and Tezuka, 1992, p. 22)

CORE TECHNOLOGIES OF A BUSINESS

All industries use several technologies in their products, production, and services. Some of these technologies are uniquely necessary to the product, production, or service systems of the industry and can be called "core" technologies for the industry. The other technologies that an industry uses will also be necessary, but may not be unique, in that other technologies may be substitutable; we will call these the "supportive" technologies for the industry.

The concept of core technologies is important because the necessary technologies particular to a product or in production will differentiate competitiveness between businesses. Lagging in a core technology will surely put a business at a serious competitive disadvantage, whereas leading in core technology will defend against competitors' encroachments.

Some of the core or supportive technologies will usually be changing at a much faster rate than others; these may be called the "strategic" technologies for the industry. The strategic technologies, which are also core technologies for the industry, can be called "pacing" technologies, for competition in the industrial structure will be technically paced by these. (See Table 4.1.)

The core technologies in the subsystems of a business enterprise that are developed and held in-house for competitiveness are proprietary core technologies of the firm. As such, proprietary core technologies can be in products, production, distribution, or information. If a core technology is acquired from outside the firm, the core technology can not provide a competitive advantage to the firm but may still be critical.

Core technologies provide significant opportunities for product differentiation and for cost reduction in products. These technologies, because they are core to the firm, should not be considered static, but rather they should represent a dynamic aspect of the enterprise.

No matter how important technologies are to a firm, however, dramatic improvement may not be possible due to the nature underlying the technol-

TABLE 4.1 Centrality of Technologies to Industry

Core technology	Necessary and unique
Supportive Technology	Necessary but not unique
Strategic Technology	Most rapidly changing technologies
Pacing Technology	Change that is defining competition

ogy or due to the fact that a substituting technology is on the horizon. Technologies change at different rates in any period of business history. The more rapidly changing technologies in a historical period provide a kind of pacing of the rates of technological change for the business. For the management of technology, these pacing technologies are the most important to watch and manage, for in these, temporary competitive advantages may be found.

Accordingly, it is useful to classify all technologies as core and supportive of an enterprise system and then designate their rates of technological change in the form of a technology-pacing matrix, as in Table 4.2.

For each business, one can list the technologies as core or supportive and fill in the matrix with the rate of change of each technology: D for a dormant technology, S for a slowly changing technology, R for a rapidly changing technology. The rapidly changing core technologies provide the pacing of the rate of technology changes in the enterprise.

One can plan technology strategy for a business around how a competitive advantage could be gained by innovation in the pacing technologies of an enterprise.

Case Study: Development of the U.S. Auto Industry

This case illustrates the impact of change in core technologies on the dynamics of a new industry. The historical context is the growth of the automobile industry in the first half of the twentieth century in the United States.

In the United States, the first automobile manufacturers came from the bicycle industry. Bicycles were a new innovation in the 1880s, made possible by three things: (1) the cheapness of steel from the earlier innovations in steel production, (2) the paving of city streets, and (3) the discovery of vulcanized rubber for tires. To invent the automobile, the idea was to put some kind of engine onto a carriage made of bicycle components, for the bicycle provided steering, gearing, and wheels. Three kinds of engines were tried: (1) steam engine and wood fuel, (2) electric motors and battery power, and (3) internal combustion engine and gasoline fuel. It is interesting that carriage manufacturers did not innovate automobiles, for they traditionally worked with wood, whereas the new bicycle manufacturers worked with steel.

The year 1896 can mark the beginning of the U.S. automobile industry, because, that year, more than one auto was produced from the same plan. J. Frank Duryea made and sold 13 cars in Springfield, Massachusetts in 1896. During the next few years, many new automobile firms were founded and a variety of auto configurations were offered (Abernathy, 1978). Races were held between the three principal configurations of automobiles—steam, electric, or gasoline-powered. In 1902, a gasoline-powered car defeated electric and steam cars at a racetrack in Chicago,

TABLE 4.2   Strategic Technology Matrix

	Business 1	Business 2
Product/Service Technologies		
Core 1		
Core 2		
•		
Supportive 1		
Supportive 2		
•		
Production Technologies		
Core 1		
Core 2		
•		
Supportive 1		
Supportive 2		
•		
Distribution Technologies		
Core 1		
Core 2		
•		
Supportive 1		
Supportive 2		
•		
Information Technologies		
Core 1		
Core 2		
•		
Supportive 1		
Supportive 2		
•		
Management Technologies		
Core 1		
Core 2		
•		
Supportive 1		
Supportive 2		

establishing the dominance of the gasoline engine. Thereafter, this engine was to become the core technology for the automobile.

In 1902, the Olds Motor Works constructed and sold 2500 small two-cylinder gasoline cars priced at $650. The next six years in the United States saw the growth of many small automobile firms selling different versions of the gasoline engine machine. The next key event in the history of the U.S. auto industry was Henry Ford's introduction of the famous

Model T. Henry Ford was producing automobiles and racing them to establish a reputation for performance for his automobiles. His cars were expensive, as were all other cars, principally for the well-to-do to purchase. But Ford had in mind a large untapped market—a car for people living on farms. Around 1900, more than half the U.S. population lived on farms. A car for farmers had to be cheap, rugged, and dependable. One of the major technical problems that Ford faced then was the heaviness and expense of the steel chassis of the automobile.

While attending a road race in 1905, Henry Ford saw a French-made automobile crash. Since Ford was very interested in what kinds of cars competitors were making, he went over to investigate the wreckage after the race. Examining the broken engine of the auto, he picked out a valve stem that seemed unusually light. He took it back to his factory to learn its composition, and found that it was made of vanadium steel (steel with the element vanadium added as an alloy). Measuring the strength of the vanadium steel sample, Ford learned that it was more than twice as strong as American steel. This was the technical breakthrough Ford was seeking: Building a car chassis from vanadium steel could weigh less and be stronger than current autos.

The strength of the chassis was critical. As an early car bumped and bounced over rough roads, its chassis could crack, wrecking the car. Also, motors were bolted directly to the chassis, and sometimes, in the violence of a pothole, a motor could be literally twisted in half. Ford said to Charles Sorensen (who helped him design the Model T): "Charlie, this means entirely new design requirements, and we can get a better, lighter, and cheaper car as a result of it" (Abernathy, 1978, p. 31).

Ford contracted with a small steel company in Canton, Ohio, to research how to produce vanadium steel. With vanadium steel for a new strong, lightweight chassis, Ford began the design of his Model T. He decided to put the engine in front and suspend it on rubber mounts, rather than bolting it directly to the chassis. Ford also chose the best ideas of the time—a magneto ignition (no batteries), drive-shaft powering of the rear wheels (no bicycle chains), and so on. Ford designed the basic form of the automobile that was to dominate for the next 50 years. The Model T became the design standard of the auto industry.

Ford's Model T was the right product at the right time for the right market at the right price. Performance, timing, market, price—these are the four factors for commercial success in innovation. Ford captured the auto market from 1908 through 1923, selling the majority of automobiles in the United States in those years.

CORE TECHNOLOGY INDUSTRIAL LIFE CYCLE

The rate of technical change of the core technologies of a product affects the dynamics of the market growth of the industry. David Ford and Chris

Ryan suggested that a chart of market volume over time for an industry would reflect the underlying maturation of the core technologies of its product, which they called a core technology "industrial life cycle" (Ford and Ryan, 1981).

Figure 4.1 shows the general pattern of growth for a new industry begun on a basic technological innovation. Market volume does not begin to grow until the application launch phase of the new core technology begins with an innovative product. (In the case of the automobile, that was in 1896, when Duryea made and sold the first 13 cars from the same design.)

The first technological phase of the industry will be one of rapid development of the new product, during the applications growth phase. (For the automobile, this lasted from 1896 to 1902, as experiments in steam, electric, and gasoline-powered cars were tried.)

When a standard design for the product occurs, rapid growth of the market continues. (For the automobile, this occurred with Ford's introduction of the Model T design.) Industrial standards ensure minimal performance, system compatibility, safety, repairability, and so forth. Sometimes, these standards are set through an industrial consortium and/or government assistance

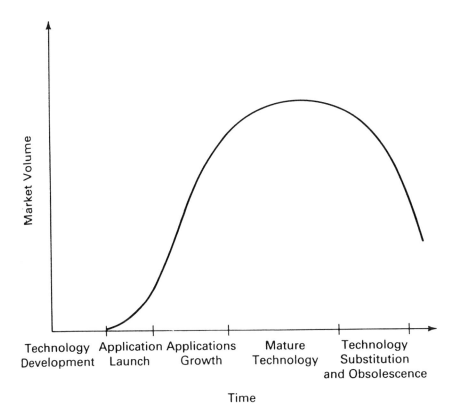

Figure 4.1. Core technology industrial life cycle.

(e.g., safety standards). But, usually, in a new technology, a performance standard emerges from a market leader.

William Abernathy and James Utterback pointed out that, usually, the pattern of early innovations in a new-technology-based industry will be, first, product innovations (improving the performance and safety of the product); later, innovations shift to improving the production process to make the product cheaper and with better production quality (Abernathy, 1978; Utterback, 1978).

Figure 4.2 plots the rate of product and process innovations over time. One sees that the rate of product innovations peaks about the time of the introduction of a design standard for the new-technology product. Thereafter, the rate of innovations to improve the product declines, and the rate of innovations to improve production increases.

This occurs because, until the product design has been standardized, a manufacturer cannot focus on improving the production processes that will produce such a design.

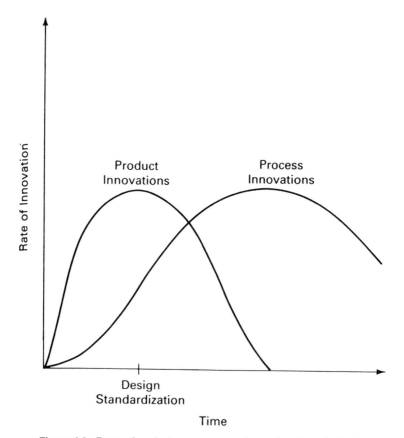

Figure 4.2. Rates of product versus process innovations in an industry.

After the key technologies of the industry mature, the market for the industry will eventually saturate. This level of market for the industry will continue, unless the key technologies for the industry become obsolete by technology substitution. Then, the market volume of the industry based on the older key-technology product will decline to zero or to a market niche.

INDUSTRIAL PRODUCT STANDARDS

Industrial product standards are critical for the growth of a large-volume market. Sometimes, these occur by a design standard set by an innovative product of a competitor, and sometimes, they occur by cooperation between industry and government to set acceptable safety or consumer standards. In this latter case, the process for establishing industrial product standards can be complex. Reaching consensual agreement between industrial firms and government about industrial standards is neither simple nor easy.

For example, Antonio Bailetti and John R. Callahan examined how public standards in communications were managed by industrial firms (Bailetti and Callahan, 1995). They pointed out that managing industrial standards was a difficult problem for several reasons, such as the need for collaboration among competitors in setting standards, the problems of complexity and uncertainty in standards when the technology is rapidly changing, and the need for international cooperation in setting standards. They suggested that a systematic procedure facilitated standard-setting, which could be built on three kinds of groups in the standards process: information management, commercial exploitation, and standard development. These are standard-making groups that focus, respectively, on information formulation, product standards setting, and maintenance and evolution of standards.

Case Study: Numbers of U.S. Auto Firms over Time

This case study illustrates the effect of competition, during a core technology industrial life cycle, on the number of firms in the industry. The historical setting is, again, the first century of the U.S. auto industry.

In 1909, in the new U.S. auto industry, there were 69 auto firms, but only half of these survived the seven years to 1916 (Abernathy, 1978). In 1918, Ford's new Model T began putting many of these out of business, as the new design standard for automobiles captured the majority of the auto market. Competitors had to quickly redesign their product offerings to meet the quality and price of the Model T. By 1923, in the United States, only eight firms succeeded in doing this and remained—with about 26 firms failing in the four years from 1918 to 1923. The eight remaining firms then were General Motors, Ford, Chrysler, American Motors, Studebaker, Hudson, Packard, and Nash.

The depression of the 1930s and the Second World War interfered with the normal growth of the auto industry in the United States, but after that war, the market growth of the U.S. auto industry resumed. The average annual sales of cars in the United States peaked around 1955, at about 55 million units sold per year. By then, General Motors had attained close to a 50 percent market share. In 1960, the number of domestic auto firms remaining was four: General Motors, Ford, Chrysler, and American Motors.

The 1970s saw the beginning of significant U.S. market share being taken by foreign auto producers. During that decade, gasoline prices jumped due to the formation of a global oil cartel. American producers did not meet the demand for fuel-efficient cars. In 1980, U.S. auto producers faced a desperate time, with obsolete models, high production costs, and low production quality.

During the 1980s, foreign companies' share of the U.S. market climbed to one-third, and there were three remaining U.S.-based auto firms: General Motors, Ford, and Chrysler.

INDUSTRIAL COMPETITION IN A CORE TECHNOLOGY INDUSTRIAL LIFE CYCLE

Figure 4.3 charts the general pattern of the number of producing firms compared to the market levels of an industrial life cycle. The numbers of firms peaks around the time of the design standardization in the industry, and declines over time to just a handful of firms. This is just the pattern we saw in the above example of the U.S. auto industry, and it is a general pattern seen historically in all national or global market-based manufacturing industries. (The pattern is not necessarily true of regionally-based service or retail industries.)

The early phase in a new manufacturing industry based on a new core technology is an exciting time, as product improvements continue and the market growth is rapid. Then, companies spring up in the new industry like weeds, started by individuals entering the industry and by employees leaving established companies and starting up their own firms. At this time, all the new firms are small and change is swift.

The product design standardization phase is the time when the rapid winnowing-out of the many new firms occurs. This happens even as the market grows dramatically, with only a few firms capturing this growth. Survival depends on gaining market share.

As the industry enters a mature technology phase, the number of domestic firms declines to a very small number. Also, in the mature technology phase, international competition becomes very important, and firms struggle globally for international markets.

The general form of the core technology life cycle and its impact on the numbers of competitors in an industry has been seen historically in many

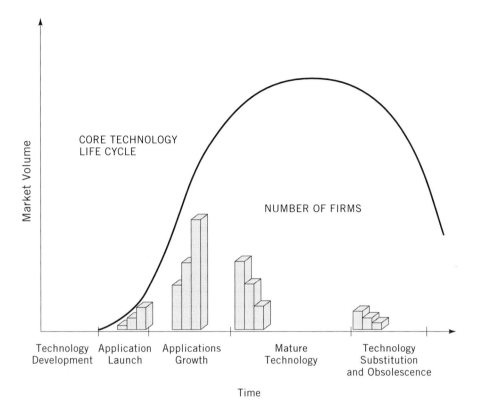

Figure 4.3. Number of firms in an industry as the core technologies mature.

industries. For example, James M. Utterback and Fernando F. Suarez charted the numbers of competitors over time in several industries: autos, TV, tubes, typewriters, transistors, supercomputers, calculators, and IC chips (Utterback and Suarez, 1993). Each of these industries showed the same pattern, with a large number of competitors entering as the new technology grew, and then the numbers peaking and dramatically declining due to intense competition as the technology progressed and matured.

For a mature technology industry, market levels remain relatively constant—since without new technology, market level is determined only by replacement rates and demographics.

Which of the many companies begun in a new industry finally succeed in emerging and surviving as dominant firms? Alfred Chandler suggested that a successful firm is the first among its competitors to make necessary investments in (1) advancing the new technologies, (2) large-scale production capacity, (3) developing a national distribution capability, and (4) developing the management talent to grow the new firm. He called such firms "first movers" (Chandler, 1990).

The reason the investment in advancing technology is important is that in the early competition, as the technology changes rapidly, a competing company must not lag behind technologically and can gain competitive advantage from being innovative.

Investments in large-scale production capacity and a national distribution system are necessary for an emerging firm to gain a dominant market share in the new national market. Also, in a global economy, a first-mover firm must also move to establish its presence in international markets.

Developing management talent to run a growing, large firm is also a necessary investment. For example, a cause of the failure of many firms to have long-term survival after an initial success has been the failure of the founder of the firm, as an entrepreneur, to build a management team that can succeed the founder.

COMMODITY PRODUCTS

As the core technologies in an industry mature, products become relatively undifferentiated in technical performance. Price and quality of the product become the primary competitive factors. Products then are often called "commodity"-type products, since they all look technically alike.

Industries are called "high-tech" industries when the core technologies in their products and production are changing rapidly. High-tech products can provide high gross margins because of the limited competition in the technical performance high-tech products provide. However, all high-tech products eventually become commodity-type products as the industrial life cycle matures. The survivors of the industrial shake-out that occurs as an industry is technologically maturing are producers:

- Who are low-cost, high-quality producers; and
- Who have established national distribution capability.

The profit margin for a firm's product lines are determined by two kinds of conditions: internal and external. Internal conditions consist of the efficiency of the firm and the strategy of management. External conditions are the balance of supply to demand in the industry. External factors include the number of competitors, how rapidly technology is changing, and how rapidly the market is growing. When an industry reaches technological maturity and products are commodity-type, then the profitability of any firm in the industry is bounded by the external conditions of the number of competitors and production capacity of the industry.

Case Study: Commodity-Type Products of the Chemical Industry in the 1980s

This case study illustrates the importance of planning in an industry around the concept of technology maturation of product lines. The historical setting for this case is a major technological turning point for the world's chemical industry as the twentieth century ended. During that century, many of the chemical industry's product lines had been high-tech and therefore high value-added (i.e., having large profit margins). By the 1970s, however, technological innovation in many areas of chemistry that were producing the ideas for new chemical product lines had slowed down. Moreover, the world chemical industry was facing the possibility of new competition from chemical plants based in the Middle East.

European and U.S. chemical firms were seeing a future chemical industry with low profitability in commodity-type products and excess world production capacity, whereas earlier they always had been accustomed to innovative and high-profit-margin products.

The chemical industry originated from manufacturers of gunpowder. For example, the largest chemical firm in the United States, E. I. duPont de Nemours, began when an emigrant French family manufactured gunpowder in the early nineteenth century. Then, in the 1850s, artificial dyes were invented using the then-new science of chemistry. This was followed by other innovations in artificial fertilizers. Then, after 1900, new innovations in artificial materials were invented, beginning with Bakelite, cellophane, neoprene, nylon, and so forth. The innovations in these areas of dyes, fertilizers, and new materials drove the technological innovation and economic expansion of the chemical industry for over a hundred years, from 1850 through 1970.

By the 1980s, however, the industry faced a problem of lacking growth from innovation. As one newspaper at that time summarized:

> It happened in steel, it happened in copper, and now it is starting to happen to basic petrochemicals. . . . A once-thriving domestic industry reaches maturity. . . . Then upstart producers in developing countries, which often have lower costs for raw materials and labor, build spanking new plants. This floods the world with excess capacity and forces many manufacturers in the developed countries to shutter their higher-cost operations. This oft-repeated trend is under way in the petrochemical industry. It has sparked a shake-out among the nation's manufacturers of basic petrochemicals such as methane and ethylene, used as building blocks for more sophisticated chemicals.
>
> —(*New York Times,* 1984)

In the late 1970s, lesser industrially developed countries with substantial oil reserves began to build petrochemical production facilities to compete in petrochemicals in the world markets. They were able to do this using production techniques for bulk chemical processes that had been invented in the period from 1880s through the 1950s. In the 1970s, these technologies were well known and widely disseminated through education and technical literature.

PRODUCT-LINE STRUCTURING OF AN INDUSTRY

The core technology industrial life cycle is still an oversimplification of the dynamics of industrial structures. An important complication results from the structuring of an industry by product lines. The applications of a new-technology-based industry often facilitates specialization in an industry around product lines.

A "product line" is a class of products embodying similar functionality and similar technology, and produced by similar production processes.

Product lines in a firm are different classes of a product produced for market segmentation. Product lines in an industry are classes of a generic product that different groups of firms in an industrial sector produce for different application systems. For example, in the automobile industry, different product lines emerged early in the industry's history according to different applications of land transportation: passenger cars, trucks, motorcycles, and tractors. After the First World War, the heavily armored military tank emerged as a fifth product line. By the 1990s, in the United States, the three domestic auto manufacturers, General Motors, Ford, and Chrysler, had both passenger car and truck divisions. The tractor industry in the United States consisted of firms that produced only tractors, farm and construction models, and general auto firms that also had tractor divisions, principally Ford. The long-term surviving tractor-producing firms in the United States were Caterpillar and International Harvester. U.S. firms solely in the truck industry did not produce passenger cars. In the United States in 1990, the surviving purely truck firms were Navistar (formerly a division of International Harvester) and Mack Trucks. On the other hand, in the Japanese auto industry, which began in the 1930s with Nissan and Toyota, the first product lines produced were trucks, and military truck production dominated the industry through World War II. After that war, Nissan and Toyota began producing passenger cars, first for the domestic market and then for the global market. In the 1960s, they innovated the small pickup product line, which grew to become a significant segment of the truck market. Historically, no automakers entered the motorcycle business in the United States, Europe, and Japan. But in Japan, some motorcycle makers, such as Honda and Suzuki, became major producers of passenger cars.

Product lines can evolve within an industry to serve different broad market niches and different broad applications, which was the case of product lines in the automobile industry. Product lines can also evolve from advancing technology. An example is the computer industry, which has organized around mainframes, minicomputers, workstations, and personal computers.

Case Study: Product-Line Lifetimes in Tire Cords

This case study illustrates that a product line can have a finite life when there is a later substitution of a technically superior product line for a prior product line. The historical setting is the first half of the twentieth century of rubber tire production for automobiles. Figure 4.4 shows an example of finite product lifetimes for tires made of different cord materials. At first, tires were produced using cotton fibers in the sidewalls of the rubber tires to reinforce the tire. These reinforcing cotton fibers were replaced by rayon fibers, which were stronger and made stronger tires. Rayon, in turn, was replaced by nylon fibers to further increase tire strength. Nylon fibers were replaced by polyester fibers for another advance in strength. Since each tire cord reinforcement material was stronger than the prior cord materials and gave stronger tires, each product-line cord was replaced by the stronger product-line cord as the price of the new cord material declined.

Here, we see that the general product line of the tire continues, but the product lines of particular types of tires from different cord materials had finite lifetimes.

PRODUCT-LINE LIFETIMES

We recall that technological change can be continuous or discontinuous; and we recall that continuous change reinforces industrial structures and

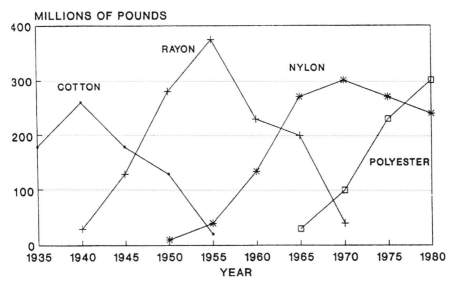

Figure 4.4. Product line lifetimes in tire cord materials.

discontinuous change creates and alters industrial structures. One form of a discontinuous technological change was the next-generation technology system. When a next generation of a technology system of a product is innovated, then the product line changes, making the earlier product line obsolete and introducing a "next-generation product-line."

Industrial structures are altered by next-generation product lines because these constitute discontinuous innovations.

> *The industrial sector that produces the product line in an industrial value chain that undergoes the discontinuous change will be altered.*

When technology changes sufficiently, product lines can undergo sufficient change as to render prior product lines obsolete. When a specific product line is replaced by a next-generation product line, this is called a "product-line lifetime."

Figure 4.5 depicts the general form of a product-line lifetime expressed as market volume of the product line over time. After the new product line is innovated, there is a rapid growth in market volume to a saturation point in the market. This saturation point occurs, as in the concept of the industrial technology life cycle, after product and process innovations in the product line have stopped, and the core technologies of the product line are mature. The market volume of the product line then remains at a level determined

VOLUME OR PERCENT OF MARKET

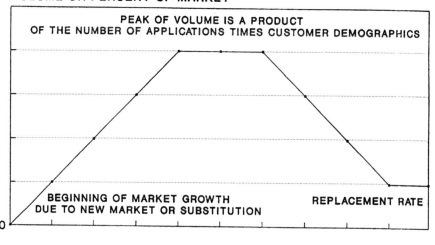

Figure 4.5. Product-line lifetime.

by replacement rates and demographics. If a new-technology-based product line begins substituting for the existing product line, then market volume for the older product line begins to decline, as the older product line has become technically obsolescent in the marketplace.

Product-line lifetimes are terminated by technical obsolescence.

OBSOLETE INDUSTRIES

An industrial sector producing a technically obsolete product line will become industrially obsolete. Thus, an industrial core-technology life cycle can become a composite of several core technology life cycles when there is a series of next-generation product lines in that industrial sector.

The societal function of an industry never becomes obsolete; however, key technology that an industry uses to serve that function may become obsolete. An industry becomes obsolete and dies only when a substituting technology develops to replace the key technologies of the existing industry. A business trying to exist in an obsolete industry will fail, unless the business helps to redefine the industry.

As examples, the ocean transportation function continues although ship construction had to be restructured in the nineteenth century from key technologies of sails to power and wood structures to iron structures. In contrast, throughout the twentieth century, the automobile industry never became obsolete, because there were no effective substitutions for the key technologies of the internal combustion engine fueled by petroleum.

Case Study: High-Tech in the Pharmaceutical Industry

How does an industry become high-tech? The answer lies in the scientific bases of the industrial core technologies. This case study illustrates the importance of science in reinvigorating an industry. The historical setting is the middle of the twentieth century, when the world's pharmaceutical industry became "high tech." The pharmaceutical industry did not begin as a high-tech industry, but was transformed into one in the second half of the twentieth century, due to advances in biological science from outside the industry.

Until the 1930s, the industry manufactured only a limited number of unpatented products and marketed without prescriptions, as summarized by Henry Gadsden (a pharmaceutical executive with the U.S. firm of Merck) in 1983:

> You could count the basic medicines (in the 1930s) on the fingers of our two hands: morphine, quinine, digitalis, insulin, codeine, aspirin, arsenicals, nitroglycerin, mercurials, and a few biologicals. Our own Sharp and

Dohme catalog did not carry a single exclusive prescription medicine. We had a broad range of fluids, ointments, and extracts, as did most firms, but we placed heavy emphasis on biological medicines as well. Most of our products were sold without a prescription. And 43 percent of the prescription medicines were compounded by the pharmacists, as compared with 1.2 percent today.

—(NAE/NRC, 1983)

This change began with important scientific discoveries made in the 1930s. Scientists discovered a series of natural products important to health—among them, vitamins and hormones. These discoveries were made in universities, supported by private foundations. These newly discovered vitamins and hormones were next manufactured by the pharmaceutical industry. They conquered age-old diseases, such as scurvy, pernicious anemia, beri-beri, pellagra, and other endocrine deficiencies.

The second contribution to the pharmaceutical industry came from medically discovered anti-infective drugs that came from compounds synthesized by the chemical industry. For example, in 1908, chemists in the laboratories of the German firm, I.G. Farben, synthesized a dye called sulfanilamide. Much later, in 1932, an I.G. Farben scientist, Gerhard Domagk, experimented with a derivative of sulfanilamide, protosil, to see if it could kill streptococcal infection. Domagk tried it on mice and rabbits, and it worked. News of this discovery reached German medical doctors. In 1933, protosil was first used in a trial on a human being, successfully treating blood poisoning from staphylococcal septicemia (Thomas, 1984). This discovery led to a series of experiments using sulfa-based chemicals for treatments of bacterial infections. This became a second high-tech product line for the pharmaceutical industry.

The third class of major pharmaceutical drugs also originated in a university. In 1928, at Oxford University in England, Alexander Fleming discovered that a bread mold could kill bacteria; he named the effective chemical compound penicillin. However, it was not until 1939 that pure penicillin was chemically isolated, by Howard Florey and Ernst Chain. The medical power of penicillin on infectious bacterial diseases was known, but industry did not immediately begin to manufacture it. This was due to the long time that had elapsed between penicillin's discovery and the demonstration of its broad and powerful therapeutic properties, so that there were no patents on penicillin. To manufacture penicillin in large quantities and in high purity would take a large amount of further production-oriented R&D, and no firm was then willing to make that investment without proprietary patent protection.

However, by 1939, the Second World War had begun, and British and American scientists urged their governments to support the development of mass production of penicillin to assist in healing wounded soldiers. In the early 1940s, the U.S. government provided $3 million to support industry and universities to research the production of penicillin. Further-more, the U.S. government provided funds to U.S. pharmaceutical firms to build plants for the wartime production of penicillin. After the war, these plants were sold to private firms for half their cost, as a reward for their participation in creating the massive production of penicillin needed for the war effort (NAE/NRC, 1983).

With these three technological bases—vitamins and hormones, sulfa drugs, and penicillin drugs—the pharmaceutical industry rapidly changed after the war ended. Its leadership saw the importance of research and the product opportunities to explore variations on sulfa and penicillin drug formulations. Pharmaceutical firms built corporate research laboratories and expanded divisional research laboratories. From the 1950s through the 1970s, pharmaceutical firms became a high-tech industry, bringing many new drugs onto the market and supporting research with very large R&D budgets.

SCIENCE BASES FOR NEW INDUSTRIAL DYNAMICS

A new high-tech industry begins from new core technologies based on new scientific advances—scientific technology. A periodic renewing of the high-tech nature of an industry depends on scientific progress that can create new core technologies for the industry. Technology discontinuities are excellent times for aggressive technology leadership.

For example, Clayton Christensen and Richard S. Rosenbloom studied technology development in the disk drive industry. They found that technol-ogy leaders had commercial success when their innovations disrupted "es-tablished trajectories of technological progress in established markets"—technology discontinuities. Such discontinuities in technology allowed at-tackers to disrupt competitive conditions for established firms (Christensen and Rosenbloom, 1995).

SUMMARY

The dynamic pattern of market growth in a new high-tech industry depends on the rate of technological innovations in the core technologies of the in-dustry's products and production. Competitive conditions in the industry are also influenced by the rate of technological progress.

The general pattern is that market growth begins for an industry when a significant product is innovated based on a new core technology. Rapid mar-ket growth begins when a "design standard" for the product is produced. Market growth continues until (1) core technologies mature and (2) indus-trial production saturates the market demographics. The number of firms entering the industry increases as the market volume grows and dramatically decreases as the core technologies mature. Firm survival depends on being a successful "first mover" in the industry.

When the core technologies mature, the once high-tech products of the industry become commodity-type products. Profit margins in the industry come under competitive pressure, and industrial leadership goes to the low-cost, high-quality producer with excellent distribution capabilities and, ultimately, to the quality/cost leader in production. The market volume levels off for the industry until a new core technology is innovated to substitute for the existing core technologies. If a substituting technology is innovated, then the current industry structure becomes obsolete (even while the market application continues).

One complication to this pattern of core technology industrial life cycle occurs when a technology discontinuity creates next-generation products in the industrial product lines. Then, industrial restructuring can occur within the industry. The high-tech revitalization of an industry depends on the progress of science that can create new core technologies for the industry.

FOR REFLECTION

Identify an industry that became historically obsolete in its key technology. When and where did it begin? When was the technology of the industry most innovative? How did the industry end? Do remnants of the industry survive today?

____5
SCIENCE AND TECHNOLOGY INFRASTRUCTURE

CENTRAL CONCEPTS

1. R&D sectors
2. Evolution of scientific bases for technology
3. Scientific technology
4. Industrial need for science
5. University/industry strategic research partnerships

CASE STUDIES

U.S. R&D Infrastructure
Origin of Biotechnology
The First Two Decades of the Biotechnology Industry

INTRODUCTION

The innovation of new basic technologies that create new industrial structures and devices arises from progress in science and technology. This progress is organizationally created in the institutional infrastructure of a nation that performs scientific and technological advances. This can be called the science and technology (S&T) infrastructure of a nation, or also, as it is sometimes called, a national research and development (R&D) system.

The first technological innovation problem in the S&T infrastructure arises from the issues of when and how industry needs scientific progress. Since industry directly uses technology, and not science, industry needs new science only indirectly, and:

1. When technological progress in an existing technology cannot be made without a deeper understanding of the science underlying the technology, or
2. When new basic technologies need to be created from new science.

The second problem relating to an S&T infrastructure arises from the fact that the university, and not industry, is the primary initiator of science progress. Thus, industry must look to the university for progress in science; however, universities traditionally have not advanced science in forms directly usable by industry, nor in a timely manner.

Accordingly, the practical issues of a nation's science and technology policy are:

1. How can firms use universities to stay technically competitive in a world of rapidly changing science and technology?
2. How can universities obtain funds from both industry and government to support the advancement of science and respond to industrial needs for science in appropriate forms and timely manners?
3. How can a government best direct its R&D support to facilitate partnerships between universities and industries?

We will review how science progresses and its impact upon technological opportunities, and we will review the organizational infrastructure for coupling scientific and technological progress.

R&D INFRASTRUCTURE

The modern means of fostering economic development and national competitiveness through technological progress are provided by the three institutional R&D sectors of universities, industries, and governments. The institutional sponsors of R&D are principally industry and the federal government (with some support by state governments and private philanthropic foundations). The institutional performers of R&D are industrial research laboratories, governmental research laboratories, and universities. Industry has become the principal producer of technological progress, and universities the principal producers of scientific progress. Government laboratories participate (in varying degrees, by country) in performing some technological and scientific progress. Government has become the major sponsor of scientific research and a major sponsor of technological development in selected areas (such as military technology).

In the second half of the twentieth century, the R&D infrastructure of nations changed dramatically due to the increased direct participation of governments in the support of research. The United States led the way as a kind of "superpower" in the cold-war political context from 1950 to 1990. The dramatic increase in governmental support of research arose from the realization of the importance of research to developing military technology that occurred during the Second World War, and also due to a widespread recognition of the importance of R&D to commercial competition.

Alexander MacLachlan emphasized the increasing importance to U.S. firms of strategic partnerships with other research sectors, such as universities and government labs (MacLachlan, 1995). In another example, Alvin K. Klevorick, Richard C. Levin, Richard R. Nelson, and Sidney G. Winter observed that the sources of new technology for industry can come from advances in science or from technological advances from other industries, universities, or government labs, and that a given industry's capability of utilizing different sources is an important factor in the innovativeness of the industry (Klevorick et al., 1995).

This increasing use of shared research is not only a U.S. phenomenon, but also a world phenomenon. For example, P. A. Hutcheson, A. W. Pearson, and D. F. Ball studied chemical process plant and equipment suppliers in the United Kingdom as providing sources of technical innovation and emphasized the importance of appropriate alliances between suppliers and users in fostering technological innovation (Hutcheson et al., 1996). They also pointed out that these alliances are improved by the inclusion of university-based research groups in the research partnerships.

Case Study: U.S. R&D Infrastructure

In the United States in the second half of the twentieth century, R&D support increased to about 2.5 percent of GNP. In 1994, about 62 percent of R&D was sponsored by the federal government, and 38 percent by industry. About 71 percent of R&D was performed by industry, 13 percent by universities, and 13 percent by governmental laboratories. Until 1965, half of the funds spent in U.S. industry on R&D came from the federal government. Afterward, industry's expenditure of its own funds on R&D grew, so that by 1995, only 18 percent of U.S. industrial R&D expenditures came from the federal government (NSB, 1996).

A major impact of the Second World War on the U.S. R&D infrastructure was to provide a massive increase in federal funding for university research. Previously, some academic research had always been connected to industry (particularly in engineering), but then the federal funds altered the balance. As N. Rosenberg and R. R. Nelson noted:

One consequence (of World War II) was a shifting of emphasis of university research from the needs of local civilian industry to problems associated with health and defense.

—(Rosenberg and Nelson, 1994, p. 338)

Together, the changes in U.S. R&D infrastructure created a kind of division of research labor between industry and universities:

R&D to improve existing products and processes became almost exclusively the province of industry, in fields where firms had strong R&D capabilities. . . . What university research most often does today is to stimulate and enhance the power of R&D done in industry, as contrasted with providing a substitute for it.

—(Rosenberg and Nelson, 1994, p. 340)

Internationally, similar patterns have emerged in the R&D infrastructures of all industrialized countries. Differing from country to country, however, are the level and emphasis of governmental R&D expenditures, modes of governmental support of R&D, roles of governmental laboratories, and research structures of universities.

For example, in the United States, Germany, Japan, France, and the United Kingdom, the ratio of R&D expenditures to GDP ranges from 2 percent to 3 percent. All other developed countries spend less, at a ratio of less than 1.5 percent. Also, only the United States, France, and United Kingdom spend government R&D funds primarily on defense. Other countries distribute government R&D support more evenly over defense, industrial development, health, energy, and space (NSB, 1996).

In the United States, the budgetary classifications of research are usually called basic, applied, or developmental. About one-eighth of R&D is for basic, one-fifth for applied, and two-thirds for developmental research (NSB, 1996). Since developmental research is closer to innovation, more funds are devoted to developmental research where the payoff is nearer; thus, most of the research performed in industry is developmental.

For example, annually in the United States, industrial research was about four-fifths development and about one-fifth applied, with a little basic research (specifically, in U.S. industry in 1995, research was 78 percent developmental, 19 percent applied, and 3 percent basic). In industry, the budgetary category of "basic research" is fairly close to what ordinarily is meant by "science," and "applied" and "developmental" research is "technology." These figures indicate that industrial research is mostly about technology and only very little about science. This contrasts to university research, which is as much about science as technology. In the United States, university research was budgetarily classified as about two-thirds basic and one-third applied. However, in university research, the terms "basic" and "applied" do not have clear meanings, since most research is performed by graduate students for doctoral theses and published by the students and faculty in research publications. As Rosenberg and Nelson noted:

> Today, except for those fields where, in effect, university work is substituting for industrial R&D (as in forest products) university research is "basic" research. However, by this we do not mean that such research is not guided by practical concerns . . . it is a gross misconception to think that if research is "basic" this means the work is not motivated by or funded because of its promise to deal with a class of practical problems.
> —(Rosenberg and Nelson, 1994, p. 340)

SCIENCE

All basic research, and much applied research, is what we normally call science. Other applied research, and all developmental research, is what we call technology research. A general definition can be expressed as follows:

"Science" is the discovery and understanding of nature.

Science is a set of activities for research about nature. The explicit goals of scientific research are (1) to *discover* new kinds and aspects of nature and (2) to *understand* nature through observation and experimentation, resulting in the development of theory. Scientific knowledge accumulated through observation and experimentation is abstracted into scientific theory and validated by further observation and experimentation. The way to conduct observation and experimentation to develop and verify theory is called the "scientific method."

It is important to note that the definition of science does *not* contain the terms "manipulation" or "human purpose." The lack of the term "manipulation" is why science is not directly useful to the economy, but is *indirectly useful* through technology, because economy needs to manipulate nature for productive activities. This also is why science is seen as "value-neutral," because it does not have "human purpose" directly as a goal.

Science is divided into many scientific disciplines, so much so that all of science is sometimes referred to as "the sciences." The major disciplines of the sciences include mathematics, physics, chemistry, biology, computer science, social sciences, environmental sciences, astronomy, and planetary science. Each of these is divided into subdisciplines and specialties. The kinds of observational or experimental techniques used in different disciplines vary, as does how theory is socially constructed and methodologically validated. But all the different techniques fall within the general concept of scientific method, and the goals of all disciplines are to discover and understand nature.

Case Study: Origin of Biotechnology

This case illustrates how the university part of an R&D infrastructure establishes the science base for the radical innovations that create new industries. The historical setting is the century of biological research from the 1870s to the 1970s, which established the science base for the new biotechnology industry began in the late twentieth century.

New industries begin upon the innovation of a radically new basic technology. The rate of occurrence of these has depended on the rate of scientific progress—which often has taken a long time. Therefore, a long-term view is necessary in science and technology policy. For example, the final scientific event in creating the science base for the new biotechnology was a critical biology experiment performed by Stanley Cohen and Herbert Boyer in 1972 that invented the technique for manipulating DNA—recombinant DNA. The scientific ideas that preceded this experiment began about one hundred years earlier. These ideas were directed toward answering a central question in science: How is life reproduced? This answer required many stages of research to be performed, including:

1. Investigating the structure of the cell
2. Isolation and chemical analysis of the cell's nucleus and DNA
3. Establishing the principles of heredity
4. Discovering the function of DNA in reproduction
5. Discovering the molecular structure of DNA
6. Deciphering the genetic code of DNA
7. Inventing recombinant DNA techniques

The Structure of the Cell. By the early part of the nineteenth century, in the then-new scientific discipline of biology, scientists were using an eighteenth-century invention, the microscope, to look at bacteria and cells. Cells are the constituent modules of living beings. Scientists saw that cells have a structure consisting of a cell membrane, a nucleus, and protoplasm, contained within the wall and surrounding the nucleus. In 1838, Christina Ehrenberg was the first to observed the division of the nucleus when a cell reproduced. In 1842, Karl Nageli observed the rodlike chromosomes within the nucleus of plant cells. Thus, by the middle of the nineteenth century, biologists had seen that the key to biological reproduction of life involved chromosomes, which divided into pairs when the nucleus of the cell split in cell reproduction (Portugal and Cohen, 1977).

Discovery and Chemical Analysis of DNA. Scientific attention next turned to investigating the chemical nature of chromosomes. Science is a system containing disciplines; the techniques and knowledge in one discipline may be used in another discipline. Chemistry and physics, as scientific disciplines, provided tools and knowledge to the scientific discipline of biology. In 1869, a chemist, Friedrich Miescher, reported the discovery of DNA, by precipitating material from the nuclear fraction of cells. He called the material nuclein. Subsequent studies showed that DNA was composed of two components—nucleic acid and protein.

While these studies were occurring, there also were continuing improvements in microscopic techniques. The more detail one wishes to observe, the more the means of observation of science need to be improved. For the microscope, specific chemicals were found that could be used to selectively stain the cell. Paul Ehrlich discovered that staining cells with the new chemically-derived coal-tar colors correlated with the chemical composition of the cell components. (This is an example of how technology contributes to science, for the new colors were products of the then-new chemical industry.)

In 1873, A. Schneider described the relationships between the chromosomes and various stages of cell division. He noted two states in the process of mitosis (the phenomenon of chromosome division, resulting in the separation of the cell nucleus into two daughter nuclei). In 1879, Walter Flemming introduced the term "chromatin" for the colored material found within the nucleus after staining. He suggested that chromatin was identical with Miescher's nuclein.

At this time, studies of nuclear division and the behavior of chromosomes were emphasizing the importance of the nucleus; but it was not yet understood how these processes were related to fertilization. In 1875, Oscar Hertwig demonstrated that fertilization was not the fusion of two cells, but the fusion of two nuclei. Meanwhile, the study of nucleic acid components was progressing. In 1879, Albrecht Kossel began publishing in the literature on nuclein. Over the next decades, he (along with Miescher) was foremost in the field of nuclein research; they and Phoebus Levine (1869–1940) finally laid the clear basis for the determination of the chemistry of nucleic acids.

As early as 1914, Emil Fisher had attempted the chemical synthesis of a nucleotide (component of nucleic acid, DNA); but real progress was not made in synthesis until 1938. Chemical synthesis of DNA was an important scientific technique necessary to understand the chemical composition of DNA. One of the problems was that DNA and RNA were not distinguished as different molecules until 1938. (This is an example of the kinds of problems that scientists often encounter, that nature is more complicated than originally thought.) By the end of the 1930s, the true molecular size of DNA had been determined. In 1949, C. E. Carter and W. Cohn found a chemical basis for the differences between RNA and DNA. By 1950, DNA was known to be a high-molecular-weight polymer with phosphate groups, linking deoxyribonucleotides between 3 and 5 positions of sugar groups. The sequence of bases in DNA was then still unknown. By 1950, the detailed chemical composition of DNA had finally been determined, but not its molecular geometry. Almost 100 years had passed between the discovery of DNA and determination of its chemical composition.

The Principles of Heredity. From 1900 to 1930, while the chemistry of DNA was being sought, the foundation of modern genetics was also being established. Understanding the nature of heredity began in the nineteenth century, with Darwin's epic work on evolution and with Mendel's pioneering work in genetics. Modern advances in genetic research began in 1910, with Thomas Morgan's group researching heredity in the fruit fly, *Drosophila melanogaster.* Morgan demonstrated the validity of Mendel's analysis and showed that mutations could be induced by X-rays, providing one means for Darwin's evolutionary mechanisms. By 1922, Morgan's group had analyzed 2000 genes on the *Drosophila* fly's four chromosomes and attempted to calculate the size of the gene. Müller showed that ultraviolet light could also induce mutations. (Later, in the 1980s, an international "human genome project" would begin, with the goal to map the entire human gene set.)

The Function of DNA in Reproduction. While the geneticists were showing the principles of heredity, the mechanism of heredity had still not been demonstrated. Was DNA the transmitter of heredity, and, if so, how?

Other scientists were studying the mechanism of the gene, with early work on bacterial reproduction coming from scientists using bacterial cultures. R. Kraus (1897), J. A. Arkwright (1920), and O. Avery and A. R. Dochez (1917) had demonstrated the secretion of toxins by bacterial cultures. This raised the question of what chemical components in the bacterium were required to produce immunological specificity. The search for the answer to this question revealed a relationship between bacterial infection and the biological activity of DNA.

Also (as early as 1892) viruses and their role in diseases were identified. In 1911, Peyton Rous discovered that rat tumor extracts contained virus particles capable of transmitting cancer between chickens. In 1933, Max Schlesinger isolated a bacteriophage that infected a specific kind of bacteria. Next, scientists learned that viruses consisted mainly of protein and DNA. In 1935, W. M. Stanley crystallized the tobacco mosaic virus, which encouraged scientists to study further the physical and chemical properties of viruses.

Meanwhile, in 1928, Frederick Griffith had shown that a mixture of killed infectious bacterial strains with live noninfectious bacteria could create a live infectious strain. In 1935, Lionel Avey showed that this transformation was due to the exchange of DNA between dead and living bacteria. This was the first clear demonstration that DNA did, in fact, carry the genetic information. By 1940, work by George Beadle and Edward Tatum further investigated the mechanisms of gene action by demonstrating that genes control the cellular production of substances by controlling the production of enzymes needed for their synthesis. The scientific stage was now set to understand the structure of DNA and how DNA's structure could transmit heredity.

Structure of DNA. We have reviewed the long and many lines of research and different disciplinary specialties necessary to discover the elements of heredity (genes and DNA) and their function (transmission of heredity functions). Before technology could use this kind of information, however, one more scientific step was necessary—understanding the structural mechanisms. This step was achieved by a group of scientists who were to be later called the "phage group" and would directly give rise to the modern scientific specialty of "molecular biology" (Judson, 1979).

In 1940, M. Delbruck, S. Luria, and A. Hershey, founders of the phage group, began collaborating on the study of viruses. One of their students was James Watson, who studied under Luria at the University of Illinois. Watson graduated in 1951 with a desire to discover the structure of DNA. He heard that the Rutherford Laboratory in Cambridge, England, was strong in the X-ray study of organic molecules and an X-ray picture of DNA would be necessary. With a postdoctoral fellowship from the United States, Watson asked Luria to arrange for him to do his postdoctoral research at the Rutherford Laboratory.

Meanwhile, the Rutherford Laboratory did have a researcher using Xrays to try to determine the structure of DNA: Rosalind Franklin, a young, bright scientist from Portugal. Franklin had come to Rutherford and begun a project in X-ray diffraction of crystallized DNA under the supervision of Maurice Wilkins, a senior scientist at the Rutherford Laboratory.

X-ray crystallography was a technique for sending X-rays (high-energy photons) through crystals and inferring the structure of the crystals from the diffraction patterns the X-rays produced from the structure. (An analogy would be to have a line of pilings near the shore and watch a big wave come in and produce smaller waves from the pilings; and, from watching the smaller waves, calculate backwards to measure the spacing between the pilings.)

Once at the Rutherford Laboratory, Watson found a collaborator in

Francis Crick, a graduate student working on a physics degree. Crick was a bit older than Watson, as his graduate studies had been interrupted by service in the Second World War. Watson brought a knowledge of biology and organic chemistry to their collaboration, and Crick a knowledge of physics—these were necessary for the job of constructing a molecular model of DNA.

But a critical piece of information they needed was a good X-ray diffraction picture of a crystal of DNA. Franklin was working on this, and it was not easy. There were two crystalline forms of DNA, and only one of these would yield a good picture. Moreover, it had to be oriented just right to get a picture that would be interpretable as to structure. Franklin finally obtained a good picture of the right form.

Meanwhile, Watson had learned of Linus Pauling's and Robert Corey's work on the structure of crystalline amino acids and small peptide proteins. From these, Pauling published a structural description of a first example of a helical form of a protein. Pauling was one of the most famous organic chemists in the world. Watson saw himself in a scientific race with Pauling to be the first to discover the structure of DNA.

Watson then conjectured: What kind of X-ray diffraction picture would a helical molecule make? If DNA were helical, Watson wanted to be prepared to interpret it from an X-ray and asked another young expert in diffraction modeling for a tutorial. He was told how, if the picture were taken "head-on" down the axis of the helix, to measure the angle of the helix. Watson was thus equipped to interpret an X-ray picture of DNA.

Watson and Crick scoured the chemical literature about DNA and had been trying to construct "ball-and-wire-cutout" models of DNA. At first they had tried a triple-helix model, but it didn't work. Finally, Watson heard that Franklin had obtained a good picture, but he feared Franklin would not show it to him. Franklin was as fierce a competitor as Watson, and was not willing to show her picture before she had time to calculate its meaning. Watson sneaked a peak at the picture without Franklin's permission. It was clearly a helix—a double helix—the tutorial for Watson on crystallography had paid off.

Quickly, Watson measured the pattern and rushed to Crick with the information on the angle of the helix. Watson and Crick put their model together in the form of a double helix, two strands of amino acid chains, twisting about each other like intertwined spiral staircases. They used the angle for the helix as measured from Franklin's picture. All the physical calculations and organic chemistry fit together in the model beautifully. This was the holy grail of biology—the double-helix structure of DNA.

The structure itself was informative. It clearly indicated the molecular action of DNA in the mitosis of cell reproduction. DNA was structured as a pair of twisted templates, complementary to one another. In reproduction, the two templates untwisted and separated from one another, providing two identical patterns for constructing proteins—that is, reproducing life. In this untwisting and chemical reproduction of proteins, life was biologically inherited.

In 1995, Watson, Crick, and Wilkins were awarded the Nobel Prize in biology. Unfortunately, Rosalind Franklin was not so honored, because of her untimely death before the prize was awarded.

Genetic Coding. By the early 1960s, it was clear that the double-helical structure of DNA was molecularly responsible for the phenomenon of heredity. Proteins serve as structural elements of a cell and as catalysts (enzymes) for metabolic processes in a cell. DNA provides the structural template for protein manufacture, replicating proteins through the intermediary templates of RNA: DNA structures the synthesis of RNA, and RNA structures the synthesis of proteins. What was not yet clear was how the information for protein manufacture was encoded in the DNA. In 1965, Marshall Nirenberg and Philip Neder deciphered the basic triplet coding of the DNA molecule. The amino acids that composed the DNA structure, acted in groups of three acids to code for a segment of protein construction.

Thus in one hundred years, science had discovered the chemical basis for heredity and understood its molecular structure and mechanistic function in transmitting heredity information.

Recombinant DNA Techniques. Several scientists began trying to cut and splice genes. In 1965, Paul Berg at Stanford planned to transfer DNA into *Escherichia coli* bacteria, using an animal virus (SV40 lambda phage). *E. coli* bacteria can live in human intestines, and the SV40 virus is a virus of monkeys that can produce tumor cells in cultures of human cells. Because of the dangerous nature of the SV40 virus, Berg decided not to proceed with the experiment, publishing a design for hybridizing bacteria in 1972. Berg organized a historic meeting on safety, the Conference on Biohazards in Cancer Research in California, on January 22–24, 1973 (Olby, 1974). This stimulated later government action to set safety standards for biotechnology research.

A colleague at the University of California responded to Berg's request for some EcoRI enzyme, which cleaves DNA (and leaves the "sticky ends" of the cut DNA). Berg gave the enzyme to one of his students, Janet Mertz, to study the enzyme's behavior in cutting DNA. Mertz notices that when the EcoRI enzyme cleaved an SV40 DNA circlet, the free ends of the resulting cut and linear DNA eventually re-formed by itself into a circle. Mertz asked a colleague at Stanford to look at the action of the enzyme under an electron microscope. They learned that any two DNA molecules exposed to EcoRI could be "recombined" to form hybrid DNA molecules. Nature had arranged DNA so that once cut, it respliced itself automatically.

Another professor in Stanford University's medical department learned of Janet Mertz's results. Stanley Cohen then also thought of constructing a hybrid DNA molecule from plasmids using the EcoRI enzyme. Plasmids are the circles of DNA that float outside the nucleus in a cell and manufacture enzymes the cell needs for its metabolism (the DNA in the nucleus of the cell are principally used for reproduction). In November 1972, Cohen attended a biology conference in Hawaii. He was a colleague of Boyer, who had given the EcoRI enzyme to Berg (and Berg's student Mertz). At a dinner one evening, Cohen proposed to Boyer that they create a hybrid DNA molecule without the help of viruses. Another colleague at that dinner, Stanley Falfkow of the University of Washington at Seattle, offered them a plasmid, RSF1010, to use; RSF1010 confers resistance to antibiotics in bacteria, so that they could see whether the recombined DNA worked in the new host.

After returning from the Hawaii conference, Boyer and Cohen began joint experiments. By the spring of 1973, they had completed three splicings of plasmid DNAs. Boyer presented the results of these experiments in June 1973 at the Gordon Research Conference on Nucleic Acids in the United States.

After one hundred years of scientific research into the nature of heredity, humanity could now begin to deliberately manipulate genetic material at a molecular level—and a new industry was born, biotechnology. Boyer and Cohen would win Nobel Prizes. Boyer would be involved in the first new biotechnology company (Genentech) to go public and would become a millionaire. The days for biologists to be industrial scientists had begun.

SCIENTIFIC BASES FOR TECHNOLOGY

Science has provided the knowledge bases for scientific technology, in this kind of pattern:

1. Scientists pursue research that asks basic, universal questions about what things exist and how things work. (In the case of genetic engineering, the science base was guided by the questions: What is life? How does life reproduce itself?)
2. To answer such questions, scientists require new instrumentation to discover and study things. (In the case of genetic research, the microscope, chemical analysis techniques, cell culture techniques, X-ray diffraction techniques, and the electron microscope were some of the important instruments required to discover and observe the gene and its functions.)
3. These studies are carried out by various disciplinary groups specializing in different instrumental and theoretical techniques: biologists, chemists, and physicists. (Even among biologists, specialists in gene

heredity research differ from specialists in viral or bacterial research.) Accordingly, science is pursued in disciplinary specialties, each seeing only one aspect of the existing thing (much like the tale of the blind philosophers who never saw a whole elephant, but went around the elephant feeling each part and puzzling how it all went together). Nature is always broader than any one discipline or disciplinary specialty.

4. Major advances in science occur when sufficient parts of the puzzling object have been discovered and observed and someone imagines how to put it all together properly (as Watson and Crick modeled the DNA molecule). A scientific model is conceptually powerful because it often shows both the structure and the dynamics of a process implied by the structure.

5. Scientific progress takes much time, patience, continuity, and expense. Instruments need to be invented and developed. Phenomena need to be discovered and studied. Phenomenal processes are complex, subtle, multileveled, and microscopic in mechanistic detail. (In the case of gene research, the instruments of the microscope and electron diffraction were critical, along with other instruments and techniques. Phenomena such as the cell structure and processes required discovery. The replication process was complex and subtle, requiring determination of a helix structure and deciphering of nature's coding.)

6. From an economic perspective, science can be viewed as a form of societal investment in the possibilities of future technologies. Since the time for scientific discovery is lengthy and science is complicated, science must be sponsored and performed as a kind of overhead function in society. Without the overhead of basic knowledge creation, technological innovation eventually stagnates for lack of new phenomenal knowledge for its inventive ideas.

7. Once science has created a new phenomenal knowledge base, inventions for a new technology may be made by either scientists or by technologists (for example, scientists invented the recombinant DNA techniques). These radical technological inventions start a new technology S-curve. This is the time to begin investment in a technological revolution and to begin new industries based on it.

8. When the new technology is pervasive across several industries (as genetic engineering is across medicine, agriculture, forestry, marine biology, materials, and so on), the technological revolution may fuel a new economic expansion. The long waves of economic history are grounded in scientific advances that create basic new industrial technologies.

9. There are general implications for management. Corporations should be supportive of university research that focuses on fundamental questions underlying core technologies of the corporation. Corporations

need to perform some active basic research in the science bases of their core technologies to maintain a "window on science" for technological forecasting.

SCIENTIFIC TECHNOLOGY

Technology uses science to know and understand nature. Science constructs mathematical models of nature by theory and experiment. These models can be used for prediction of technical performance when nature is manipulated. By predicting technical performance, the engineer can design (prescribe) the degree of performance required for an application of the technology. Thus, both scientific theory and observation/experimentation are useful to technology.

Scientific methodology has turned out to be critical to the invention of new technologies and to the systematic improvement of technologies. The power of the scientific mechanistic perspective enables technologists to understand and predict the phenomena underlying a technology. Predicting the phenomena underlying a technology, in turn, enables technologists to prescribe technical performance of the technology.

Thus, in modern times, all the major new technologies on earth have been invented based on scientific progress. This fact gives rise to the second core concept in MOT (in addition to technological innovation) which is "scientific technology":

> *"Scientific technology" is a manipulation and use of nature for human purpose, based on recognized scientific phenomena.*

Technology has existed since humans became tool makers, but scientific technology needed to wait for the advent of science. The dramatic change from the "ancient" world to what we call the "modern" world was due to the rise of scientific technology, requiring: (1) the origin of science, (2) the application to economic production of new technological inventions that science made possible, and (3) continuing dynamic interaction between the new science and new technology.

Case Study: The First Two Decades of the Biotechnology Industry

This case study follows the early years of the biotechnology industry and illustrates how technological and commercial progress depends on the complexity of nature underlying the technology. In 1996, a technical journal summarized the condition of the industry:

> Fighting waves of hype and pessimism—while trying to create products and access markets—tests [biotechnology] firms' ability to endure.
> —(Thayer, 1996, p. 13)

By 1996, the biotechnology industry had created 35 major therapeutic products, which then had total annual sales of more than $7 billion. These biopharmaceutical products were used to treat cancer, multiple sclerosis, anemia, growth deficiencies, diabetes, AIDS, hepatitis, heart attack, hemophilia, cystic fibrosis, and some rare genetic deceases. But the industry was not initially as successful as early investors had hoped.

For example, Genentech (Boyer's firm) was the first biotechnology firm in the United States to go public. Genentech raised $30 million in a public offering in 1981 in the U.S. market. But by 1996, Genentech was not, as earlier hoped, a large pharmaceutical firm. It was profitable, but with a majority ownership by an older, larger pharmaceutical firm. Genentech's major product was TPA, a protein to control abnormal blood clotting, with total sales of $300 million and 75 percent of the market for controlling blood clotting:

> Despite TPA's success today, it took the 20-year-old company many years and many millions of dollars to prove that it had an important product.
> —(Thayer, 1996, p. 13).

Genentech's first product was a human growth hormone in 1985. Genentech had hoped that the TPA product innovation would catapult it into large-firm status, but the costs of developing and proving products and the relatively small market for TPA put Genentech into a financial crisis in 1990. To survive, Genentech sold 60 percent of its equity to Hoffman-La Roche:

> William D. Young, executive vice president of Genentech, says Roche "supported us when we needed it and provided cash and good stability for the stock price. That has allowed us to go a long way toward doing what we are supposed to do: invent, develop, manufacture, and market new products."
> —(Thayer, 1996, p. 15)

Genentech illustrates the rough road to commercial success that all the new firms in biotechnology go through. For example, some of the other startups from the early 1980s were firms such as Biogen, Amgen, Chiron, Genetics Institute, Genzyme, and Immunex. Biogen did pioneering work on proteins such as alpha-interferon and insulin. But to support itself, Biogen licensed its discoveries so that other pharmaceutical firms (such as Schering-Plough, Hoffmann-La Roche, Eli Lilly, SmithKline Beecham, and Merck) could market their products.

In the 1990s, most of the marketing of the new biotechnology therapeutic products were through these older established pharmaceutical firms, rather than the new biotechnology firms that pioneered pharmaceutical recombinant DNA technology. Genentech had partnered with Hoff-man-La Roche, and Chiron with Ciba-Geigy. Genetics Institute and Im-munex were majority-owned by American Home Products. The exception to this pattern of the fate of early biotechnology firms was Amgen, which, in 1995, had become an industry leader in biotechnology and an indepen-dent, fully integrated biopharmaceutical producer with sales of $1.82 bil-lion.

Why had the early hoped-for big profits in biotechnology not occurred, although biotechnology has survived and continues to develop? The an-swers to this were in science, in technology, and in economics:

> Early expectations, in hindsight considered naive, were that drugs based on natural proteins would be easier and faster to develop. . . . However, . . . biology was more complex than anticipated.
>
> —(Thayer, 1996, p. 17)

For example, one of the first natural proteins, alpha interferon, took ten years to be useful in antiviral therapy. When interferon was first produced, there had not been enough available to really understand its biological functions. The production of alpha interferon in quantity through biotech-nology techniques allowed studies and experiments to learn how to begin to use it therapeutically. This kind of combination—developing the tech-nologies to produce therapeutic proteins in quantity and to use them thera-peutically—took a long time and many developmental dollars.

Cetus spent millions of dollars and bet everything on interleukin-2 as an anti-cancer drug, but failed to obtain the U.S. Food and Drug Adminis-tration's (FDA) approval to market for this purpose. Subsequently, in 1992, Chiron acquired Cetus. Even in 1995, interleukin-2, which was eventually approved by the FDA, earned for Chiron only 4 percent of its revenues of $1.1 billion. About this, George B. Rathmann commented:

> [T]he pain of trying to get interleukin-2 through the clinic just about bank-rupted Cetus and never has generated significant sales.
>
> —(Thayer, 1996, p. 17)

The innovation process for biotechnology industry in the United Sates included (1) developing a product, (2) developing a production process, (3) testing the product for therapeutic purposes, (4) proving to the FDA that the product was useful and safe, and (5) marketing the product. In fact, recombinant DNA techniques were only a small part of the technol-ogy needed by the biotechnology industry and the smallest part of its innovation expenditures. The testing part of the innovation process to gain FDA approval took the longest time (typically seven years) and the great-est cost.

Because of this long and expensive FDA approval process in the United States, extensive partnering continued to occur between U.S. biotech firms and the larger, established pharmaceutical firms. For example, in 1995, pharmaceutical companies spent $3.5 billion to acquire biotechology companies and $1.6 billion on R&D licensing agreements (Abelson, 1996). Also, pharmaceutical firms spent more than $700 million to obtain access to data banks on the human genome that was being developed by nine biotechnology firms.

The U.S. government role in supporting science was essential to the U.S. biotechnology industry, as noted by Henri Termeer, chairman and CEO of Genzyme and chairman of the U.S. Biotechnology Industry Organization:

> The government has a very big role to play [in helping] to decrease the costs. Support of basic research through NIH [National Institutes of Health] is very important to continue the flow of technology platforms on which new breakthrough developments can be based.
>
> —(Thayer, 1996, p. 19)

In this case study of the early decades of the biotechnology industry, we see that the scientific importance of understanding the molecular nature of biology (the discipline now called "molecular biology") had proved to be the future of the pharmaceutical industry, as an essential methodology to develop new drugs. Yet the making of money from the technology of recombinant DNA had been harder and longer than expected. The reason was that science turned out to be more complicated than anticipated. Nevertheless, the biotechnology industry technology depended on, and continues to depend on, new science. In turn, the technology needs of the biotechnology industry have helped drive the discoveries in biological science.

The progress of a new technology depends on the progress of understanding the complexity of nature underlying the technology.

The funding going into the biotechnology industry has turned out to be worth the expense, for biology continues to be the future of the pharmaceutical industry. As George Rathmann accurately summed up:

> It doesn't have to follow that science automatically translates into great practical results, but so far the hallmark of biotechnology is very good science and that now is stronger and broader than ever. . . . The power of the science is ample justification that there should be good things ahead for biotechnology.
>
> —(Thayer, 1996, p. 18)

INDUSTRY/UNIVERSITY RESEARCH COOPERATION

The effective use of universities and other sources of research partnerships depends on managing the partnerships. For example, Ronald S. Jonash emphasized the importance of managing the processes for obtaining new technology from sources external to the firm (Jonash, 1996). For science, in particular, the management question boils down to how best to perform university-based science for industry.

One answer to this lies in the concept of a next-generation technology. For example, through support of industry/university research cooperation that the U.S. National Science Foundation has provided over the last two decades, it has identified an effective concept for coupling university science to industrial technology needs: the idea of university and industrial research projects around the vision of a next-generation technology (Betz, 1997).

University basic research cannot be planned for a radically new technology, but it can be planned for a next-generation technology (NGT) system. This is an appropriate goal for universities and industries to perform together. Incremental innovation is an inappropriate goal for university research, because industrial research is best positioned for this.

After a basic invention, research can be planned by focusing on the generic technology system and its production processes and their underlying physical phenomena. NGT research advances a previously existing technology and thus can be planned.

NGT focused and targeted basic research can be planned for:

1. Generic technology systems and subsystems for product systems
2. Generic technology systems and subsystems for production systems
3. Physical phenomena underlying the technology systems and subsystems for product systems and production systems

Since science can be focused on the physical phenomena underlying technologies, "targeted basic research" can be scientific research motivated by any of the above. University and industry strategic partnerships for research on next-generation technologies are an excellent way for university science to be planned for industrial needs.

SUMMARY

Research in the three institutional sectors of universities, industries, and governments are the modern means of advancing science and technology for national economic development and competitiveness. During the twentieth century, this modern form of a national S&T infrastructure has evolved and continues to evolve. Modern universities facilitate industrial innovation through advancing the sciences that underlie modern technologies and the basic engineering techniques used by industry. University research sometimes invents new basic technologies and explores the knowledge bases for next-generation technology systems.

Scientific progress takes much time, patience, continuity, and expense. Instruments need to be invented and developed. Phenomena need to be discovered and studied. From an economic perspective, science can be viewed as societal investment in the possibilities of future technologies.

When industry needs science, it can do the research itself at great expense, high risk, and trouble, or industry can turn to universities to do the science. University and industry strategic partnerships for research on next-generation technologies are an excellent way for university science to be planned for industrial needs.

FOR REFLECTION

The computer and software industry began in the second half of the twentieth century. Identify the basic innovations that laid the basis for these industries. Who innovated them? Why were these innovations economically pervasive?

____6
INNOVATION LOGIC

CENTRAL CONCEPTS

1. Macro-level of innovation processes
2. Micro-level of innovation processes
3. Linear and circular innovation processes
4. Culture and roles in innovative organizations

CASE STUDIES

The 1976 Japanese MITI LSI Chip Project
Japan Captures the World Memory Chip Market

INTRODUCTION

Within a science and technology infrastructure, activities are carried out to create technological innovation. What kind of activities are pursued and how they are ordered consititute a kind of logic for proceeding systematically about technological innovation. This logic has been called an "innovation logic," and procedures organized around an innovation logic have been called an "innovation process."

An innovation process covers the intellectual grounds from the origin of basic knowledge about nature (science), to the invention of means to manipulate nature (technology), to the design of products, services, and processes that utilize such inventions for commercial purposes (economy). The logical steps necessary to beneficially make these interactions provide the basis for an innovation process. In order to manage technological change, it is useful to have an overall picture of innovation processes, the procedures by which technological innovation occurs in a society and in a firm.

We will address the following issues:

- What are the different levels of innovation processes in a society?
- What are the logical stages in these processes?
- What kinds of organizational cultures and roles facilitate innovation processes in an organization?

MACRO-LEVEL OF INNOVATION PROCESSES

There are three macro levels of innovation: international, national, and industrial levels. The international level of innovation involves the sharing of knowledge and cooperation in creating knowledge across nations and throughout the world. The national level involves sharing of knowledge across the institutional sectors of a nation. The industrial level involves sharing of knowledge across the firms in an industry.

Case Study: The 1976 Japanese MITI LSI Chip Project

This case study illustrates an innovation process that occurred at a national level and was intended to create an internationally competitive industry. Its historical context is the major international industrial restructuring in the consumer electronics industry that began in the 1960s and 1970s, with Japanese firms emerging as world leaders by 1980.

The three basic innovations in electronics were (1) the electronic vacuum tube, invented in 1910; (2) the transistor, invented in 1951; and (3) the semiconductor integrated circuit (IC) chip, invented in 1959. The IC chip replaced electronic circuits composed of transistors and other components wired together.

Through the 1960s and 1970s, new U.S. firms established to produce IC chips dominated the world market. In the 1970s, however, Japanese electronics firms emerged as the world leaders in consumer electronics, innovating the transistorized pocket radio, transistorized television sets, transistorized hi-fi, and home VCRs. The original innovations occurred in the late 1950s and 1960s, when transistor technology began to substitute for electron vacuum tube technology in electronic circuits. Although the U.S. invented and innovated the transistor and provided leadership in innovating transistors into computers and into defense electronics, it was the Japanese firms that took the leadership in innovating transistors into the consumer electronics of radios, hi-fidelity audio, and television in the 1970s.

This technological substitution of transistors for tubes in consumer electronic products continued to provide commercial opportunities for the growth of Japanese electronics firms, as U.S. electronics firms lagged behind and were slow to make this change. When the 1970s began, the U.S. consumer electronic firms were beginning to die, with Japanese imports taking over the U.S. market.

This extraordinary success in consumer electronics stimulated the Japanese government to wish to advance technology for its industry. In 1970, Japanese electronics firms did not yet produce IC chips, only transistors. These firms and the Japanese government decided that national production of IC chips was essential if they were to continue their growth.

The history of technical progress in IC chips changed not incrementally up a traditional technology S-curve, but in a series of technology S-curves, each only in the exponential portion. The invention of the IC chip in 1959 put many transistors on a chip. By the mid-1960s, the technology had improved to put hundreds of transistors on a chip. By the end of the 1960s, the technology had improved again, to putting on thousands of transistors on a chip (MSI). In the middle of the 1970s, the technology would jump again, to put tens of thousands of transistors on a chip (LSI).

In 1970, the Japanese government sponsored a large-scale integration (LSI) next-generation technology project in Japan (Ypsilanti, 1985). The LSI scale for which the MITI project aimed was intended to bring Japanese electronics firms into current technical competitiveness with U.S. chip firms. This decision by a nation to enter a new technology at the next-generation level of the technology (rather than the current generation) was, at the time, an unusual decision for a developing economy. In the past, most countries with developing economies had transferred technology in the form of the current generation of capability from countries with more advanced economies. This, however, posed the problem of leaving the developing economy still behind competitively, as an advanced economy was then proceeding on to a next generation of the technology. The MITI decision of the Japanese government, while more risky at the time, had the advantage of putting the firms in Japan at the cutting edge of international technology in IC chips.

MITI provided the equivalent value of $10 million in research grants to Japanese electronics firms from 1970 to 1975. Some of the grants went to electronics firms producing transistors and consumer electronics devices, and some to firms producing production equipment. Japanese firms engaged in the project registered abut 1000 patents on chip technologies.

The research projects were aimed at inventing and understanding new processes to refine the production process of chips to increase transistor density—i.e., to make the transistors smaller. To make an IC chip, first a circuit is designed and photographed. Next, a very small image of the photograph is projected onto a silicon disk to a photoresist coated on the silicon. This "picture" of the circuit is etched in the silicon by a physical process, such as an acid or deposition of a reactive chemical. Other materials, such as aluminum, need to be deposited on the silicon to form conductors; and the silicon layers for the transistors need to be "doped" by injecting donor atoms, impurities, into the silicon material. The steps of photographic projection of a circuit feature and etching are repeated until a circuit object, such as a transistor, is constructed.

The first electronics product to use LSI technology capability was the 16K memory chip for computers (16,000-bit storage in dynamic random access memories (DRAM)). In 1976, this 16K memory chip was innovated by U.S. firms and by Japanese firms—thanks to the MITI-sponsored project. By 1979, the demand for the 16K chip was so high that U.S. firms could not satisfy the demand in the United States. Japanese firms satisfied their own internal national demand but also exported to the United States, to gain 20 percent of that market.

U.S. computer manufacturers were surprised to find that the Japanese 16K memory chips were of higher production quality than U.S. produced memory chips. At that time, about 20 percent of U.S.-produced chips were defective, whereas fewer than 10 percent of Japanese-produced chips were defective. So, by 1980, the Japanese electronics firms were fully competitive with the U.S. in memory chip technology and even more advanced in manufacturing practice.

In this case study, we see an example of planning a technological innovation at a national level, a shared R&D project between government and national firms, to create a national technical capability.

SCIENCE AND TECHNOLOGY INFORMATION TRACKS

As we saw in a modern R&D infrastructure, research in industry is focused primarily on advancing technology, whereas research in universities is focused primarily on advancing either science or generic technology. Published knowledge from these research endeavors follows different information tracks: one for science and one for technology. Figure 6.1 summarizes how these two research tracks interact in creating knowledge for innovation (modified from Haeffner, 1980).

In the science track, the current state of scientific knowledge is archived in scholarly journals of disciplinary-focused scientific and professional societies. Textbooks summarize and codify this knowledge and communicate it to the next generation of scientists, engineers, and other professionals through education courses at the undergraduate level, graduate level, and continuing education.

From an understanding of parts of the current state of scientific knowledge, researchers in scientific, engineering, and professional disciplines pose

SCIENCE TRACK

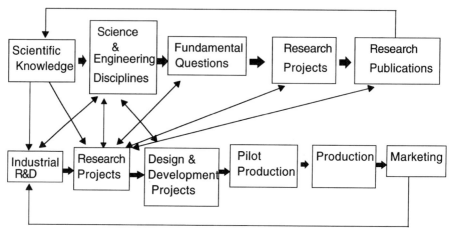

TECHNOLOGY TRACK

Figure 6.1. Science and technology research tracks.

fundamental questions to advance the state of knowledge in their disciplinary specialties. These fundamental questions are framed as research projects, with prescribed methods to obtain answers. From successful research projects come research publications in scholarly journals, which add to the previous scientific knowledge.

The research publications are put into an "open literature" of international science, so that all basic knowledge is available to everyone—all competitors, all nations, and future generations. This is the reason why all technological knowledge eventually diffuses throughout the world—because the science bases of technology are in the public domain. No nation or firm can hope to keep an intellectual monopoly on knowledge for a long period of time. Only temporary legal monopolies are possible for technology, because of patent laws; but this is always a limited period, usually 17 years.

This is an important point: It makes the knowledge base of technology essentially open over the long term. This is why industrialization has been open to all nations.

New technology arises in the technology track from invention occurring in research projects in this track. In the technology track, industrial R&D strategy provides the basis for focusing and funding most technology research projects. These research projects use information from the current state of scientific knowledge, science and engineering and professional disciplines, and scientific research publications. The arrows connecting the science and technology research tracks indicate these knowledge contributions of science to technology. Textbooks and handbooks from the general state of scientific knowledge and from particular disciplines provide the underlying knowledge base of facts, theories, instrumentation, and methods that are used in the technology research projects. The latest results in scientific progress are found in the science research publications.

Successful technology research projects do not usually result in research publications, but in inventions that are further developed in design and development projects. Technology projects do not usually result in publications, because technology is a private good in improved products, production, or services. Novel and useful new ideas in technology are sometimes published in the form of a patent, because the patent provides temporary proprietary rights. Technology development projects are aimed at new or improved products, services, or production. As appropriate, new products or services need to be produced at a pilot plant level and tested before initial production and marketing begin.

The goal of the science research track is to increase public knowledge about nature, whereas the goal of the technology research track is to increase private economic benefits. There are many intellectual interactions between these two tracks. The science and engineering disciplines take many problems for research from industrial research projects. And industrial research projects take much information from science and engineering disciplines and from scientific research publications.

Case Study: Japan Captures the World Memory Chip Market

This case study illustrates, at the micro-level of innovation processes, how the fruits of an national innovation process can be exploited by firms to dramatically alter a competitive structure. The historical context follows of the previous case, six years later. In 1980, the state of technology in the semiconductor chip industry both in the United States and in Japan, was aiming at a next generation of technology, then called very-large-scale integration (VLSI). VLSI was to provide millions of transistors on a single chip.

The first product to embody this technological capability was the 64K DRAM chip. From the previous case, we saw that technological parity between U.S. firms and Japanese firms had occurred in 1975, when the 16K memory chip was introduced with LSI technology. In 1981, the race was on to be the first to innovate the 64K memory chip. The first to the market would likely grab the majority of the market. (When a market is, developed and waiting for a next-generation product, the first to innovate becomes the largest market-share holder.)

American firms designed a 64K chip with some refined features over the 16K chip design, to produce a smaller-sized chip. In contrast, Japanese firms took a standard 16K chip design (innovated earlier by the U.S. firm Mostek) and simply enlarged the design to the 64K level, for they understood the importance of being the first to market. This resulted in a slightly larger sized chip, but this made little difference to the customer. What mattered to the customer was who produced the chip first, and with the highest quality. That race went to the Japanese electronic firms.

In November 1981, Japanese electronic firms introduced the 64K memory chip, and U.S. firms followed nine months later. Those nine months were critical. By the end of 1982, Japanese electronic firms held 76 percent of the memory chip market, and the American firms would never recover. The U.S. chip industry simply stopped mass-producing memory chips (with a few exceptions).

Why did one timing slip at a technology discontinuity (from the 16K chip to the 64K chip) permanently lose the memory chip market for the United States? The reasons were (1) the increasing cost of developing next generations of technology for chip production and (2) the structure of the consumer electronics industries to bear and recover the costs. IC chip technology and next-generation technology products requiring increased transistor density demanded ever more expensive production facilities as the technology refined. When chip density increased by an order of magnitude, the cost of building a chip processing facility to produce that density also increased by about an order of magnitude. Unless a chip producer had a substantial market share, there was no cash flow to finance the investment for the next-generation production facility. When the U.S. chip-producing firms lost market share, they lost the cash flow to finance the development of the next generation of technological production capability. Once out of the race for memory chip markets, U.S. firms were out of that race for good. It is a sobering thought for a global economy.

MICRO-LEVEL OF INNOVATION PROCESSES

The micro-level of innovation processes occurs at the level of individual firms innovating new technologies in their products and production. There, the logic of technological innovation begins with *invention* and moves into development and design. The first goal of the early research and development after invention is to stage a technology-feasibility demonstration of the invention to show that the invention does work.

The next logical step is to improve the working of the invention enough to show it can perform in an application; this result is displayed as a *functional prototype*. The logical step after this is to improve the working of the invention further, to show that it has the features and safety and size to work in a product or service; this is called the *engineering prototype*. The next logical step is to design the invention-embedded product, process, or service for a salable good or service; this is called an *engineering design*. The last logical step is to redesign the product, process, or service into a form that can be produced in volume at quality and cost targets; this is called a *manufacturing design.*

1. *A technology-feasibility demonstration of an invention shows that the invention works.*
2. *A functional prototype shows that the invention performs well enough for a market application.*
3. *An engineering prototype shows that an invention has the features, safety, and size to be designed as a product for the market application.*
4. *An engineering design embeds the invention into a designed product, process, or service that can be sold into the market.*
5. *A manufacturing design redesigns the product/process/service for production in volume and at quality and cost targets.*

Next, a production system needs to be developed and designed to produce the product, process, or service.

A prototype production process shows a production system that can produce the product, process, or service at volume, through put, quality, and cost targets.

Then, the investment must be made in constructing the production process and producing a large enough initial inventory of the product to begin selling it.

Learning to produce a new product in large volumes at high quality and to reduce cost is a continuing process after the initial innovation.

Thus, technological innovation begins with invention of a new product concept, but then proceeds in stages of developing the product and production. This is why technological innovation is expensive—all these steps occur after invention. The problem of managing technology thus requires two activities: encouraging invention and managing successful innovation.

LINEAR AND CYCLIC INNOVATION PROCESSES

We recall that technological innovation can be classified by the size of the innovative step—incremental or discontinuous. There are two general forms for the innovation process:

1. A cyclic innovation process for incremental innovations
2. A linear innovation process for radical innovations (basic or next-generation) (Gomory and Schmitt, 1988)

The cyclic innovation process for incremental innovations occurs in the product development cycle of firms. The linear process for discontinuous innovations occurs in corporate research laboratories, in universities, in government laboratories, and in new high-tech startup companies.

The cyclic innovation process in the firm results in incremental technological innovations, for which the firm performs research and development projects aimed toward improving the performance of current product lines, services, or production processes. The results of these projects are then innovated in improved product models, new releases of service software, or improvements to production. The cyclic process adds technological improvements to existing technologies. This process has traditionally been called the "product development cycle." It is cyclic in nature because all products have finite lifetimes. More recently, this procedure has also been called the "product realization process," in order to emphasize the importance of computer-based tools in engineering design and development. Thus, incremental innovation in products occurs in a firm's product realization process.

The linear innovation process in the firm results in new business ventures, for which research projects (in corporate research laboratories or in universities) have invented new products or services. A new firm may organize around the new products, and venture capital is raised to start the firm. The linear process creates new technologies from new science.

CULTURE AND ROLES IN INNOVATIVE ORGANIZATIONS

The logic of innovation needs to be institutionalized in the culture and roles in an organization.

Culture

An organizational culture that encourages and supports innovative activity improves the chance of commercially successful innovation. Alan Frohman summarized three aspects of the corporate culture that facilitate the utilization of technology: (1) top management orientation, (2) project selection criteria, and (3) innovative systems (Frohman, 1982).

The support of top management for innovation is especially important, for top management controls the resources and rewards within the organization. In supporting R&D, management should have a clear set of technology and business strategies, which provide criteria for selection of R&D projects. The organization must also have the formal information systems and structure for managing both the invention of new technology and implementation of the technology into new products, processes, or services.

Another student of innovation, James Brian Quinn, also emphasized the importance of top executives in providing vision and creating the culture to facilitate innovation: "Visions, vigorously supported, are not 'management fluff,' but have many practical implications; they attract quality people to the company and give focus to their creative and entrepreneurial drives" (Quinn, 1985, p. 78). Quinn stressed that innovative organizations must be oriented toward markets in their innovation efforts, and should also support multiple project approaches to cover technological risk. Innovation succeeds only when market needs are satisfied through improving quality or lowering cost. Innovative organizations must have cultures that foster both technological strength and marketing strength. Since technological development is risky, multiple approaches to technology can sometimes reduce the risk, particularly as time to market is important.

Roles

What kind of roles do personnel play in innovation processes? The concept of roles in innovation processes was first proposed in a study called SAP-PHO, in which four kind of roles were then identified as essential to industrial innovation: the technical innovator, the business innovator, the chief executive, and the product champion (SAPPHO, 1972). Other researchers followed up on the idea of roles in innovation and elaborated the kinds of roles needed. For example, J. Smith and others (Smith et al., 1984) studied a sample of 10 innovations at one firm, Union Carbide, and refined the list of roles to:

1. The scientific gatekeeper
2. The inventor
3. The process/product/service champion
4. The R&D strategist
5. The R&D sponsor
6. The project manager
7. The problem solver
8. The business sponsor
9. The process user gatekeeper
10. The product user gatekeeper
11. The quality controller
12. The top management

In modern technological progress, major changes in technology tend to come directly from scientific progress. Therefore, an individual in the innovation system who maintains intellectual currency in the latest in relevant science progress is an important source of information about what may become possible in new technology. This is why corporate research laboratories employ scientists to perform basic research in selected areas of science relevant to their technologies. This is frequently called "windows on science."

The role of the inventor is key in the innovation process to the creation of new technology. Sometimes, scientists invent new technology, but, more often, engineers invent new technology.

Since, after an invention is made, it must be developed into technological improvement in products, processes, or services, the next role in the innovation process is a product/process/service champion, who pushes to create an R&D project to develop the invention. In any organization, change requires a champion to advocate and bring about the change.

In the innovation process, advocacy for technological change occurs within the R&D strategy of the firm. So the product/process/service champion must be an advocate with the people who set R&D strategy; this role influences what is innovated.

In a small organization, several of these roles will be played by the same individual. In a large organization, different individuals may play the different roles. It is important for an organization to recognize, encourage, and appropriately reward the different roles in the innovation process, if technological change is to be encouraged.

SUMMARY

Activities within an S&T infrastructure are interconnected through logic needed to accomplish technological innovation. The three macro-levels of innovation are international, national, and industrial levels. At the micro level of innovation in a firm, innovation logic has several stages, beginning with research and invention and proceeding to market introduction. Incremental innovation and basic innovation proceed with different logics, a circular production cycle and a linear innovation logic.

The logic of innovation needs to be institutionalized in the structure, roles, and culture of an organization. Corporate culture that facilitates the utilization of technology includes proper top management orientation, proper project selection criteria, and proper procedures. Important innovative roles include technical innovator, business innovator, product champions, and innovation-supportive executives.

FOR REFLECTION

Find, either in the literature or in a firm, an example of a new product or new process innovation. Write a brief case study of the innovation, identifying the steps. Also identify the key personnel in the project and describe their roles.

___7
TECHNOLOGY IN BUSINESS

CENTRAL CONCEPTS

1. Enterprise system
2. Value-adding operations
3. Subsystems of an enterprise system
4. Types of technologies used in a business
5. High-tech and commodity-type products
6. Comparing technologies for manufacturing and for service

CASE STUDIES

Sewell's Innovation of Plastic Beverage Bottles
Rediscovering Economic Value Added (EVA) in 1993

INTRODUCTION

We have been looking at the larger perspective on technological innovation, historically and at the national level. Now, we focus on the micro-level of innovation, the level of the firm. We need to understand how technological innovation fits into and establishes the productive basis of a firm. We will address the following issues:

- How does technology contribute to the enterprise system of a business?
- What kinds of technology are used in a business?
- How should management pay attention to technological changes within a business?

Case Study: Sewell's Innovation of Plastic Beverage Bottles

This case provides an illustration of the force of technological change in altering a competitive situation—enabling a small company to leap into national rank as a major company. The historical context is a major change in the container materials for bottled beverages, which occurred in the early 1980s. Before 1982, large bottles of soda were sold in heavy glass bottles, and milk was sold only in quart sizes in either glass or waxed-paper cartons. Now, large-sized beverages are all sold in plastic bottles. Charles K. Sewell of the Dorsey Corporation was the first U.S manager to foresee that technical change and exploit it, gaining a major competitive advantage.

In 1975, Sewell was manager of Dorsey's Sewell Plastics Division (which Sewell had founded earlier and sold to Dorsey). Dorsey had two other divisions, Chattanooga Glass and Dorsey Trailers. Chattanooga Glass made green Coca-Cola bottles. Sewell Plastics produced plastic containers. The Chattanooga Glass Division that dominated Dorsey's businesses, providing 60 percent of total sales (Boyer, 1983).

About this time, a chemist at DuPont had invented a new kind of plastic, called polyethylene terphthalate (PET). He found that plastic bottles made of PET were much lighter than glass and could hold (as glass did) carbonated beverages. DuPont engineers made a 2–liter beverage container out of PET with an idea of substituting this for glass bottles. By 1977, DuPont received Federal Food and Drug Administration approval to use the two-liter PET bottle as a beverage container. A cooperating machine tool company, Cincinnati Milacron, built equipment that could mass produce the PET bottle. (The fact that a chemical firm like DuPont invented both a new material and a new application for the material was not unusual; it was a deliberate technology strategy in DuPont's research to try to create new materials and applications for companies to which it supplied materials.)

Even with this dramatic innovation (a next-generation technology), most companies in 1977 making glass bottles for beverage producers ignored it. They didn't wish to make the major investments needed to change technologies.

But Sewell saw the opportunity. He saw the PET bottle technology as a way to compete with the giant glass-bottle-making companies (Owen-Illinois, Continental Group and Amoco). He went to the president and board of Dorsey, telling them of the commercial opportunity and the required investment. They approved. Sewell next ordered $4 million worth of the new equipment from Cincinnati Milacron, following soon with another $9 million order. He installed the new machines in each of his plants to serve local markets. Then he solicited beverage customers to convert to the new technology. Using plastic bottles would save them weight in transporting beverages, and their customers would prefer the lighter and unbreakable bottles. His customers responded. By 1982, Sewell Plastics had aggressively moved into the position of leader in the then $800 million two-liter container market. It was a story of a David beating out the Goliaths with innovative and aggressive management.

It is important to note that Sewell's correct business strategy had been based on two kinds of understanding: technology and market. Because Sewell was in the plastics business, he quickly understood the commercial opportunity of the PET invention. And because Dorsey's other division, Chattanooga Glass, was in the glass soft-drink container business, Sewell was able to see the potential market for the technology.

The giants in glass bottle manufacturing were slow to respond to the new technical opportunity and lost market share. But they also had to change eventually or die. In a next-generation technology change, there is no choice. When a new technology replaces an old technology, it is better to change: "The alternative is uninviting—a strong position in a flagging industry and no position at all in the successor that's emerging" (Boyer, 1983, p. 176).

ECONOMIC VALUE ADDED

The economic measure of contribution that a new technology makes to a business is how much *value* it adds to the business enterprise. This concept of "economic value added" has two subconcepts: (1) the value added to resources that a business transforms into a product for a customer, and (2) the profit obtained in the marketplace of the product's sales price to the customer over the costs of producing the product.

Thus "value-added" can be viewed from the perspective of a firm's customers or managers. From the perspective of the customer, the economic value added is the functionality and performance the product provides for the customer's applications of the product. From the perspective of management, the economic value added is the gross margin of the product.

Technology is essential to both meanings of "value-added" in that technology provides the knowledge bases for the value-adding transformations of the business and directly affects profitability. In the management of technology, the fundamental way to look at these systemic transformations that create economic value-addedness is to see the organization as an "enterprise system."

ENTERPRISE SYSTEM

Organizations are goal-directed, and create productive transformations to reach these goals. For a business, this goal-seeking has been called the "concept of the enterprise," and its productive transformations constitute the "enterprise system."

Michael Porter emphasized the importance of viewing a business as a transforming open system—a "value chain" (Porter, 1985). The value-adding activity of a business is a sequence of transforming operations, a chain of activities, which add value. The business takes resources from the economy, transforms them into products, and sells them back into the economy—adding economic value to the original resources (see Figure 7.1).

This concept of "value-adding" is also the same concept in organization theory that all productive organizations can be conceived of as open systems (Betz and Mitroff, 1974). Any organization can be seen as receiving inputs from its environment, transforming these into outputs, and sending the outputs back into the environment. As a transforming open system, a business (1) acquires material, capital, and personnel resources from the economy; (2) transforms these into goods and/or services; and (3) sells the goods and/ or services into the markets of the economy.

Since the measure of performance of business transformation is its profitability, or economic value added, the accounting system of the organization should be capable of measuring its "economic value-adding."

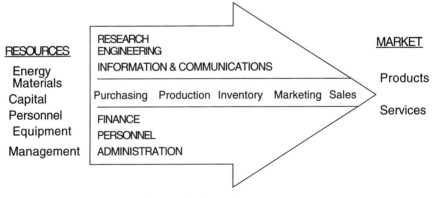

Figure 7.1. Enterprise system.

The role of technology in the value-adding activities of business is to provide the knowledge bases for the transforming functions of the enterprise.

Case Study: Rediscovering Economic Value Added (EVA) in 1993

This case illustrates that, in practice, it is not easy to measure the value adding activities in an enterprise system. This depends on the sophistication of the accounting system. Measuring the contributions of the functions of a business to the value-adding transformations is one of the important problems of modern accounting. The historical context of this case is a time in the United States in the early 1990s that followed a major financial restructuring of U.S. industry. During the 1980s, more than one quarter of large U.S. firms changed ownership.

The interests between ownership of a company (shareholders) and control of the company (management) coincide when companies are small and owners few. However, this may not be true of large firms with widely distributed stock ownership, or where large blocks of stock are held by pension or mutual funds. In the 1930s, students of industrial organization had already begun to realize that a large public company whose stocks are widely distributed, without any owner having a significantly large share, will be controlled by the top management through its control of shareholder proxies. One of the major problems of modern management is how to make management's and shareholder's self-interests more congruent. This case study is example of an attempt to improve management accountability to shareholders through improved accounting practices.

In the United States, by the 1990s, the practices of remunerating executives in large firms had often become divorced from accountability, due to the capability of management to control proxies and appoint boards of directors. Several CEOs had led some large, diversified corporations into decline, while continuing to claim large salaries and bonuses. Finally, some CEOs of underperforming major firms were fired by their boards of directors. As a popular business journal reported:

> The King is dead: Booted bosses, ornery owners and beefed-up boards reflect a historic shift in corporate power. The imperial CEO has had his day—long live the shareholders.

> Has it come to this? Consider: General Motors Chairman and CEO Robert Stempel unceremoniously unhorsed. Ditto Kenneth Olsen, founder, president, and principal executive officer of Digital Equipment. . . . The former chiefs of Compaq, Goodyear, Hartmarx, Imcera, Tenneco—the list goes on—getting in more golf than they expected at this age . . .

What's happening is a reversal of the decades-old tendency of corporate power to gather in the hands of executive officers rather than directors or owners. . . .

—(Stewart, 1993, p. 34)

Because of the influence of institutional shareholders, many U.S. CEOs began to pay more attention to the issue of "what they had done recently for their shareholders." How could this be measured? It could be measured as the recent "economic value added" (EVA) by operations:

What if you could look at almost any business operation and see immediately whether it was becoming more valuable or less? What if you as a manager could use this measure to make sure your operation—however large or small—was increasing in value? What if you as an investor could use it to spot stocks that were far likelier than most to rise high? There is such a measure . . . economic value added, or EVA. It is today's [1993] hottest financial idea and getting hotter.

—(Tully, 1993, p. 38)

Of course, in management practice, old ideas may become new ideas again :

Understand that while EVA is easily today's [1993] leading idea . . . it is far from the newest. On the contrary: Earning more than the cost of capital is about the oldest idea in enterprise.

—(Tully, 1993, p. 39)

What was new was that some corporate accounting practices began to be able to measure the total economic value-adding transformation of the enterprise. A measure of EVA requires an accounting system to record the true cost of capital against all capital employed in operations. Traditional accounting principles, up to 1993, had left many true capital costs unrecorded. For example, accounting systems at that time had usually recorded the costs of borrowed capital (interest paid) but not that of equity capital. In practice, equity capital had been listed on the books as 'paid-in capital' of sold shares. This is not a current value, however, but an historical value. In contrast, the interest paid on borrowed capital is a current, and not an historical, value.

The true cost of equity is what shareholders could be getting in price appreciation and dividends if they invested in a portfolio of companies similar in risk to the firm.

Because of outdated accounting principles, most firms could not measure their capital cost:

> Incredibly, most corporate groups, divisions, and departments have no idea how much capital they tie up or what it costs.
> —(Tully, 1993, p. 38)

In a firm, when management measures something and uses that measurement to judge performance, then personnel in the firm will respond to the measurements. For example, at CSX, CEO John Snow began using EVA:

> Snow has lots of capital to worry about, a mammoth fleet of locomotives, containers, and railcars. His stiffest challenge came in the fast-growing but low-margin CSX Intermodal business, where trains speed freight to waiting trucks or cargo ships. Figuring in all its capital costs, Intermodal lost $70 million (EVA) in 1988. Snow issued an ultimatum: Get that EVA up to breakeven by 1993 or be sold. Freight volume has since swelled 25 percent, yet the number of containers and trailers—representing a lot of capital—has dropped . . .
> —(Tully, 1993, p. 39)

ECONOMIC VALUE ADDED AND ENTERPRISE SYSTEMS

To use EVA to control the operations of a business, one needs to conceive of a business as an enterprise system. Accounting practices should be refined to measure all activities in terms of value addition, measuring the total operational activities of an enterprise.

So why does this concept of EVA have to be periodically rediscovered in business history? The answer probably is that it is easy to forget the totality of a system when one is preoccupied daily with only one part of the system. Accordingly, it is important always to remind oneself that, in total, any business should always been seen as a value-adding enterprise system.

SUBSYSTEMS OF AN ENTERPRISE SYSTEM

We recall that the range of technologies covered by MOT can be compacted into three kinds of technologies: product, production, and information. To analyze the subsystems of an enterprise system, it is now useful to decompact these three technologies into five enterprise subsystems. Product technologies are embedded in the enterprise's product system, and production technologies in the enterprise's production system. But information technologies become embedded in three distinct subsystems of the enterprise: distribution, communications, and management.

Thus, the five principal subsystems that need to be constructed and connected for an operating enterprise system are:

1. Product system
2. Production system
3. Distribution system
4. Communications system
5. Management system

A product system in a business enterprise is the conceptual architecture of a product line that a business designs to sell to a customer, either an economic good or an economic service. A product should be thought of as a system because the value-adding properties of a product or service to a customer derive from the functionality that is accomplished by the transformational system of the product

A production system in a business enterprise is the conceptual framework of the sequence of processes that a business uses to produce products. Production systems can produce either goods or services. The value-adding properties of production derive from the material transformations taking resources into products.

A distribution system in a business enterprise is the conceptual architecture of the channels by means of which information about a product line and access to purchasing products is provided by the business to customers. The value-adding properties of distribution provide access by the customer to information and acquisition of a business's products (through transforming the information state of a customer and through physically transforming the location of goods and services from the business's locations to the customer's locations). The technologies underlying a distribution system provide the functional capabilities of information and material access from the business to the customer.

A communication system in a business enterprise is the conceptual framework of the means and network for communications to facilitate decisions and coordination of the operations of the enterprise. The value-adding properties of the communications system provide coordination of the activities of the operations of a business (through transforming the matching and pacing of activities of personnel of a business and their interactions with suppliers and customers). The technologies underlying a communications system provide the functional capabilities of communicating information among the firm's personnel, suppliers and customers.

A management system in a business enterprise is the conceptual architecture of the planning and control procedures for guiding, monitoring, and evaluating operations. Management procedures in a business should be thought of as a system because the decisions and control in management procedures alter the flow of activities in the transformations of the enterprise system. The technologies underlying management procedures provide the functional capabilities to assist planning, monitoring, and evaluation of business activities.

TYPES OF BUSINESS TECHNOLOGIES

As with the subsystems of an enterprise system, the technologies a business will use can be classified into:

1. Product technologies
2. Production technologies
3. Distribution technologies
4. Information technologies
5. Management technologies

Since there are two types of products a business can provide to customers—goods and services—there will be different technologies for "product-goods" and for "product-services." The technologies for designing a "product-good" will be of two kinds:

1. A core technology for the major transforming activity of the product-good
2. Supporting technologies for the secondary features of the product-good

For example, the core-technology for an automobile product is the engine system, and the supporting technologies are for the body, chassis, and control subsystems.

A "product-service" will also require several types of technologies:

1. Arranging for the product-service
2. Delivery of the product-service
3. Scheduling of the product-service
4. Maintenance of product-service devices

For example, an airline service requires a reservation system, a ticketing system, airport receiving systems, a baggage handling system, airplanes, flight crews, an airplane fueling and maintenance system, scheduling systems—and all the technologies that go into each of these systems.

Producing a product also requires several kinds of technologies:

1. Unit production processes
2. Integration and control of unit processes into a production system

Each unit production process will use a core technology—for the major transforming activity of the process—and secondary technologies. The integrating production technologies will also require several technologies for moving material around the system and for control of the system.

For example, producing an automobile requires production systems for producing the motor, chassis, body, and the many parts that go into each. A simple part, like a gear, will require a forging process to produce the rough shape of the part and a milling process to finish the part and the setup and takedown of the part from the forging and milling equipment.

A distribution system requires several technologies:

1. Informing the customer about the functionality and availability of the product or service
2. Providing access for the customer to the product or service
3. Delivering the product or service
4. Maintenance and repair of the product or service

For example, a retail food chain requires technologies for shipping food products from manufacturer to wholesaler to retailer, for advertising food products to customers, for unpacking and shelving food products, for charging for and transporting food products to the customer's location, and for reordering food products.

A communications system uses several technologies for facilitating organizational coordination:

1. Communication types
2. Communication networks
3. Communication protocols
4. Network management

For example, communication types include telephone, radio, satellite, and computer. Modern communication technologies are converging on a digital format, so that voice, graphics, video, and data can all be transmitted in the same media.

A management system uses several tools for planning, supervising, controlling, and evaluating organizational activities. It includes information and decision aides, such as:

1. Engineering design systems
2. Sales and inventory systems

3. Purchasing systems
4. Personnel information systems
5. Accounting and financial systems
6. Manufacturing control systems

For example, the engineering function of an organization uses computer-aided design (CAD) systems; the production function uses computer-aided manufacturing (CAM) systems; finance uses computer-based accounting and management information systems (MIS); personnel use computer-based personnel record and payment systems. Communications and coordination among the functions of the organization use e-mail, local-area networks, and the like. All these systems use computer, information, communication, and engineering based technologies.

A multitude of technologies are used in any business, and technical progress should be managed in all to provide competitive advantages to the firm.

COMPARING TECHNOLOGIES FOR MANUFACTURING AND SERVICE INDUSTRIES

The typology of business technologies makes it easy to compare the differences in technologies essential to manufacturing and service firms.

Technologies for a manufacturing firm can be classified according to their use, as:

1. Technologies for a product
2. Technologies for production of products
3. Technologies for products distribution
4. Technologies for product maintenance and repair
5. Technologies for assisting the customer's applications of the products
6. Technologies for communicating and conducting transactions with customers and suppliers
7. Technologies for controlling the activities of the manufacturing firm

For a service firm, technologies can be classified as to their use as:

1. Technologies for devices used in service delivery
2. Technologies for supply and maintenance of devices used in service delivery
3. Technologies for service delivery
4. Technologies for service development
5. Technologies for assisting the customer's applications of services

6. Technologies for communicating and conducting transactions with customers and suppliers
7. Technologies for controlling the activities of the service firm

MANAGEMENT AND TECHNOLOGY

Previously, the role of technology in business had been primarily focused on product and production technologies—the knowledge and skill for designing and producing the products that the firm sells. This was reserved as the domain of the engineers. In contrast, the information and coordination to run the business was the domain of the managers. Business managers did not worry about how a product was designed or produced, as long as they got a product they could sell at a good margin. But technological change—especially the innovation of the computer—has dramatically altered this older division between management and engineering.

The new technologies in information and communication have become equally important to business as the older power and material technologies. In fact, "managing with technology" has become the newly important and fast-paced area of technological change. It is the new technology of the computer that is profoundly changing the nature of business organization and procedures.

With a broad historical brush, one can paint two major industrial revolutions in the world:

1. The first major industrial revolution, beginning in the second half of the eighteenth century, was based on the technologies of power and materials—specifically, the steam engine and means to shape and move materials.
2. A second major industrial revolution, beginning in the second half of the twentieth century, is based on the technologies of information— the computer and communications.

Previously, the jobs of designing, producing, and selling products were entirely separable responsibilities of different business functions. Now, *integration* over the different organizational functions is necessary for responsiveness of a business to changing economic opportunities and threats.

SUMMARY

Businesses should be viewed as enterprise systems, transforming open systems that add value. A business needs to have an accounting system that measures value addition in all activities of the firm.

The technologies a business will use can be classified into product, production, distribution, information, and management technologies.

FOR REFLECTION

Think of a business and identify the core technologies for the internally produced or used product lines, production systems, and service aids. Which are rapidly changing? What kind of competitive advantage could be gained from innovation in the rapidly changing technologies?

─────8
TECHNOLOGY AND COMPETITION

CENTRAL CONCEPTS

1. Technology push and market pull
2. Technical and commercial risks in technological innovation
3. Marketing experimentation
4. Economic evaluation of new technology
5. Technology diffusion and substitution

CASE STUDIES

Laser and Fiber-Optic Communications Innovations
Edison's High-Tech Business Failure

INTRODUCTION

Competitiveness is the heart of economies, making economic activities rational through adjusting prices between supply and demand. Competitors struggle with each other for market share, profits, and dominance. Thus, it is essential for the management of technology to understand exactly how technology affects competitiveness. Competitiveness is always changing, as, for example, Michael Porter has expressed:

As emphasized many decades ago, competition is profoundly dynamic in character. The nature of economic competition is not "equilibrium" but a perpetual state of change. Improvement and innovation in an industry are never-ending processes, not a single, once-and-for-all event.

—(Porter, 1990, p. 40)

Although different technologies provide the productive knowledge bases of a business, technology itself does not become a competitive factor until it changes. The reason for this is that when technology is stable, all competitors use the same technology. This occurs because science and technological knowledge is eventually open to and usable by anyone. Thus, when technologies are stable, management can delegate responsibility for these technologies to the engineering function of the firm.

When technology is changing, however, then both management and engineering need to work together strategically. Then, technological change directly affects the competitiveness of the firm. Getting management and engineering to work together strategically to exploit technological innovation for competitive advantage turns out not to be easy. Historically, only a few firms at any time appear to be successful in gaining a positive competitive advantage from technological change. We will address the following issues:

- What are the kinds of commercial risks involved in technological innovation?
- How does timing affect the risks in technological innovation?

TECHNOLOGY PUSH AND MARKET PULL

Technological innovations have been motivated either by conceiving a technological opportunity or by perceiving a market need, called, respectively, "technology push" or "market pull." Technology push occurs in cases of technological innovation that were motivated primarily by exploring the manipulation of nature in physical phenomena. Market pull occurs in cases of technological innovation that were motivated primarily by market need. Both sources of innovation have been important in the history of technology.

Studies of technological innovation used to contain arguments about which source of innovation ideas was most likely to result in a commercial success. For example, in the 1970s, Eric von Hippel studied samples of innovations in the scientific instrumentation industry: gas chromatography, nuclear magnetic resonance spectrometry, ultraviolet absorption spectrophotometry, and transmission electron microscopy (von Hippel, 1976). He found that the scientist-users of the instruments made 80 percent of the innovations in his sample, and the manufacturers of the instruments made 20 percent. From these and other studies (e.g. Von Hippel, 1976 and Von Hippel, 1982), Von Hippel argued that market pull was a more important source of innovation than technology push. This conclusion about the importance of market pull as a source of innovation follows from cases in which the user of the innovation is technically sophisticated and has research capability. (Historically, research instrumentation has been invented mostly by scientists.)

A contrasting study was performed by K. D. Knight, who looked at innovations in computers and found that manufacturers (rather than users) dominated innovation from 1944 to 1962 (Knight, 1963). In computers during this period, industry was more technologically advanced than were most computer users; and, in this case, technology push was a more important source of innovation than market pull. Other studies, such as those by A. Berger (1975) and J. Boyden (1976) of plastics innovation, wherein manufacturers were technologically sophisticated and had research capability, also documented the importance of technology push.

The conclusion is that both manufacturers and users can be sources of innovation, provided each is technologically sophisticated and can perform research. It is the locus of sophisticated technical performance that determines whether market pull or technology push will be most important for innovation in an industry.

But whatever the source, for technological innovation to be commercially successful, it must eventually create or match markets to technological possibility. Christopher Freeman summarized this nicely: "Perhaps the highest level generalization that it is safe to make about technological innovation is that it must involve synthesis of some kind of [market] need with some kind of technical possibility" (Freeman, 1974, p. 193).

Market-pull innovations most often stimulate incremental innovations, since an established market inspires the need. In contrast, radical innovations are more often brought forth as technology push seeking market applications.

Case Study: Laser and Fiber-Optic Communications Innovations

This case study illustrates technology push as a source of innovation. The historical setting is in the United States just after the end of the Second World War, when recent military technologies were continuing to stimulate new science and new invention.

The laser was invented by Charles Townes, motivated by his technological experience in wartime. Townes was a physicist who had worked in radar research during the war. He invented the amplifying device for microwave frequencies, which he called "maser"—Microwave Amplification by Stimulated Electromagnetic Radiation. The maser used the physical phenomena of pumping the electrons in atoms to higher orbits, and then when the right frequency of microwaves entered the stimulated region, these electrons fell back into their lower energy orbits giving off radiation of the same frequency as the triggering microwave signal— thus, in effect, amplifying the signal.

In 1951, Townes was a professor at Columbia University and was consulting for Bell Laboratories of the U.S. firm of AT&T. Townes was thinking about ways to produce coherent radio waves at the higher electromagnetic frequencies of visible light.

Townes' idea was to fill a quantum state of electrons of the atoms of a material by stimulating the atoms by passage of electrical fields through them. This was the same kind of physical phenomenon he had used in his invention of the maser. But this time he used the phenomenon for a different purpose: to generate monochromatic (single-frequency) light (rather than to amplify a microwave signal). When the stimulated electrons fell down into lower-energy orbits, they would all radiate light at the same frequency in the visible light region. Townes called this idea "Light Amplification Stimulated Emission Radiation," or laser (Townes, 1984).

A colleague at Bell Labs, Arthur Schalow, suggested that the laser light could be collected by putting mirrors at each end of the laser chamber. In December 1958, Townes and Schalow published the first paper on lasers.

Townes received a Nobel Prize in physics for his inventive idea. The first physicist to make an actual laser was Theodore Maiman in 1960, then working at Hughes Aircraft Company. Lasers were put to many uses and created a small industry of laser producers. The single-frequency characteristic of laser-produced light was very useful, as it could beam energy and information a long way with little spread of the spot of light. This made it useful for communications, ranging, or cutting and fusing materials.

RISKS AND VARIABLES IN SUCCESSFUL INNOVATION

All new technology-based business ventures are fraught with risks—both technical and commercial.

The technical risks arise from uncertainties about what a new technology can really do, how well it does it, what resources it consumes, how dependable it is, how easy it is to maintain and repair, and how safe it is. In other words, technical risks relate to a new technology's:

- Functionality
- Performance
- Efficiency
- Dependability
- Maintainability and repairability
- Safety and environment

The functionality of a product, process, or service refers to the kind of purposes for which it can be used. For example, different industries are often classified by purpose: food, transportation, clothing, energy, health, education, recreation, and so forth. The goods and services within these industries satisfy these different purposes. Furthermore, within a purpose there are usually different applications. For example, in transportation, there are applications of travel for business, vacation, and personal travel.

The *performance* of a good or service for a function denotes the degree of fulfillment of the product's purpose. For example, different food groups provide different kinds and levels of nutritional requirements.

The *efficiency* of a good or service for a level of performance of a function relates to the amounts of resources that are consumed to provide a unit level of performance. For example, different automobiles attain different fuel efficiencies at the same speed.

Dependability, maintainability, and *repairability* indicate how frequently a product or service will perform when required and how easily it can be serviced for maintenance and repair.

Safety has both immediate and long-term requirements: safety in performance and safety from aftereffects over time. The environmental impact of a product or service includes the impacts on the environment from production, use, and disposal.

In contrast to technical risks, commercial risks arise from uncertainties about who the customers for the technology are, how they will use the technology, what specifications are necessary for use, how customers will gain information and access to the product, and what price they will pay. These are risks related to:

- Customer type
- Application
- Specifications
- Distribution and advertising
- Price

The customer, application, and specifications together define the market niche of a product, process, or service. Distribution and advertising together define the marketing of the product, process, or service.

The price set for a new product, process, or service needs to (1) be acceptable to the market and (2) provide a large enough gross margin to provide an adequate return on the investment required to innovate the new product, process, or service.

Solving the technical variables correctly is both necessary and costly (in research and development costs). Even if successfully accomplished, the commercial variables must be correctly solved (with the production and marketing costs). Moreover, in technological innovation, initially, the two sets of variables are indeterminate (not necessarily precisely coupled beforehand). There is always a range of technical variables possible in the design of a new product, process, or service. Which variables will turn out to map correctly to the future required set of commercial variables is never clear initially, but is clarified only in retrospect.

One might think that one should start from the commercial variables and design the technical variables to match. This is, of course, the correct answer—but only when the technology is stable and not changing. The more radical the technological innovation, however, the less the existing set of commercial variables will be correct in the future.

Case Study: Thomas Edison's High-Tech Business Failure

This case illustrates the risks—technical and commercial—that occur in technological innovation. The historical context of this case study is a period in the life of one of America's most prolific and famous inventors, Thomas Alva Edison.

Edison was commercially successful with many of his inventions, but he was basically an inventor, and not a businessperson. He was interested in the commercialization of his inventions because of their importance first, their technical challenges second, and the material rewards third. Many inventors think this way. Consequently, while Edison became wealthy, he never created a business empire that survived him. Others did that from the businesses he started:

> For all Edison's creative and commercial success and his prodigious production of inventions, he never earned as much of a fortune as others might have. He fell short as a businessman and spent much of his earnings in legal battles, protecting his patents and fending off unscrupulous competitors.
> —(Peterson, 1991, p. 9)

Edison obtained 1093 patents, including ones improving the telegraph and stock ticker and inventing the mimeograph and electric locomotive—in addition to inventing the electric light, innovating the first electrical power system, and inventing the phonograph. In 1889, he was an internationally known figure, with many of his inventions being major attractions at that summer's Universal Exposition in Paris. He took a triumphal tour of Europe and was received by heads of state.

Yet, ironically, even at that time, Edison had already begun a final technological scheme that would fail and leave him deeply in debt. His plan was to revolutionize the iron-making business. He wished to become a head of that industry, as once he had been in the electrical field. The 1880s had, in fact, been difficult for Edison. He had many legal and financial problems, with his electrical empire absorbing his time. And as others moved in to take the business helm, Edison became bored. He turned back to earlier ideas he had about ore separation.

Back in 1879, Edison had investigated mining and mineral processing techniques, from platinum to gold to iron. He devised a magnetic ore separation technique for sands containing iron grains in the form of magnetite, a magnetic iron oxide, and patented it in 1880. In 1887, Edison applied for five new patents improving the concept, spelling out his ideas later in an article in the magazine *Iron Age*. The reaction by mining experts to his plans were that they was too ambitious, too radical, and too expensive. Equipment needed to be massive to pulverize quantities of stone containing iron, and feeding low-grade ores to a separator might be too expensive. A follow-up editorial in the magazine called his idea "Edison's Folly."

In 1889, Edison reorganized the Edison Ore-Milling Company he had formed six years earlier (and which had foundered). He offered his personal signature on any future notes, loans, and debts of the company. He invested $1 million of his own. Earlier, an investing and operating partner, Walter S. Mallory, had built a full-scale separator using Edison's new patents. It failed to function, and Mallory was broke in two years. Now, Edison gave Mallory employment as superintendent of his mining company, and they forged ahead. Edison and Mallory found an iron deposit on a mountaintop in Sussex County, New Jersey. In 1889, Edison's company bought the land and built a mill and separator and a mill town, Ogdensburg. Edison became a regular boarder in Ogdensburg's new hotel, supervising the construction of the separators.

The magnetic separators were the technical key to the entire process. This process started with quarrying rock with the world's then-largest steam shovel. It scooped low-to-medium-grade ore and dropped it into hoppers, which took it to the mill. In the mill, chunks of rock were carried to the top of a tall crushing house and dropped into the tower. Inside were rollers six feet in diameter and covered with sharp teeth. The rollers spun at 700 revolutions per minute, crushing boulders as heavy as nine tons in four tiers of rollers, until pieces only a half-inch in diameter emerged at the bottom.

These pieces went back up to the top of another crushing tower, with smooth rollers three feet in diameter, which crushed the rocks to sand. Next, this sand was lifted again to a separating tower, where, as it fell, 480 electromagnets tugged at the falling iron granules so that they fell to one side of a partition at the bottom. This magnetic separation was repeated two more times in two more separating towers, with the goal of the final batch of magnetic powders intended to be 70 percent pure iron. Edison intended to sell the rock residue to contractors as construction sand.

In the fall of 1891, Edison first tested the process in actual operations:

> It didn't work. The mill and its machinery simply weren't up to the task. The crushing rollers were not heavy or strong enough. . . . Ore dust penetrated everything. The bulky steel conveyor system constantly jammed. . . .

Some bearings were so overloaded that they failed within hours, despite constant oiling.

—(Peterson, 1991, p. 12)

Edison worked on the problems, but the mill was down most of the time. Compounding these problems were safety disasters. A lost iron mine was rediscovered when the blacksmith shop suddenly plummeted 80 feet into an abandoned shaft upon which it had inadvertently been built, killing three men. Two months later, more men were killed accidentally in constructing a new millhouse. Even Mallory and Edison were nearly killed when they entered the bottom of a screening tower to see why the material above was clogged—and 16 tons of ore dust fell on them, temporarily burying them.

Another problem was the ore dust itself. The steel company refused to buy any more of it:

What doesn't blow away in the railroad cars," Henry Frick of Carnegie Steel wrote to Edison, "blows out the chimney as soon as the furnace is fired.

—(Peterson, 1991, p. 12)

Edison solved this problem, however, finding a glue to cement the dust into briquettes and baking them in an oven. Finally in 1899, the mill opened for business. But, unfortunately, the commercial timing was wrong. The year before, in 1898, two brothers made the biggest iron ore discovery in the United States in the Mesabi Range of northern Minnesota:

[H]undreds of square miles of the purest iron ore in the world; it needed only to be scooped from the ground and sent directly to the furnaces.

—(Peterson, 1991, p. 13)

With the opening of the Mesabi Range, U.S. ore prices plummeted. In 1899, it was shipping ore at $3.00 a ton, delivered. Edison, at his Ogdensburg mill, could produce it only at $4.75 a ton, undelivered. In the fall of 1899, Edison's mill shut down. Edison reopened it in 1900, at his own expense, to produce enough ore to fulfill a previous contract with Bethlehem Iron—but at the Minnesota prices. During these two years, Edison accumulated personal debts that took years to pay off.

Edison and Mallory went on to use the rollers and furnaces to make cement—and to revolutionize the cement business. This enterprise it made enough money to nearly repay the $3 million that Ogdensburg had cost Edison. After this, Edison turned back to inventing—beginning to develop a new battery, the alkaline storage battery. Eventually, all the technologies that Edison invented in his iron-ore process did find commercial use:

> Edison had anticipated metal coating with his magnetic techniques, Portland cement with his crushing machinery, and taconite [iron] processing with his separation and briquetting techniques.
>
> —(Peterson, 1991, p. 14)

Successful innovation requires that the risks from both new technology and commercial application of the technology be overcome. Although Edison struggled to overcome both risks, he just had bad luck in the timing of the find of the massive new iron ore field. But this is why business is an adventure—requiring innovation, skill, courage, sagacity, financing, and, finally, some luck.

MARKET EXPERIMENTATION

Establishing the specifications for a new product or service requires correct focus on a market. The more radical the technological innovation in the new product, however, the more difficult this becomes. The reason that anticipating markets for new technologies is difficult is that radically new technologies and their applications are developed interactively. As new technology develops, new applications are often discovered, as well as new relationships to existing technologies and applications. Successful product innovation of new-generation technology requires that management recognize its significance and correctly focus the marketing of the new product lines. This is often hard to do, because new technology will often affect new markets in unenvisioned ways.

Marketing traditional expertise lies in knowing in detail the nature of existing markets. Marketing has well developed techniques for analyzing existing markets, under the term "marketing analysis". Traditional market analysis for new products approaches the task by market segmentation— identifying the customer group (industry or consumer) on which to focus the new product or service. This can be done because existing products have designed the function, applications and needs of the product in the market. Accordingly, when technological innovation is only incremental, and not a dramatically change in a product-line functionality, quality, or price, then market analysis can provide a good guide to how the innovation will be accepted or demanded by an existing market.

As technological change begins to change markets, however, then analysis of a market before such change often yields the wrong kind of guidance as to what the market will become after the change. In fact, the history of radically new products has often demonstrated that the largest market has turned out to be different from that envisioned by the innovator. For example, Kodak's plastic photographic film was intended for a professional photographer's market (which, at the time, used glass plates for negatives), but the plastic film quickly found a larger market in amateur photography, a new market the new film created.

> *The more radical the technological innovation for a market, the more wrong "market analysis" techniques will likely be about the nature of the market for the technological innovation.*

Rather than "market analysis," some have suggested that marketing should take a "learning" approach to the making of new markets through technological innovation. For example, Shanklin called the making of new markets by new technology a kind of "supply-side marketing," by which he meant "any instance when a product can create a market" (Shanklin, 1983).

Because trying to anticipate the market for radically new technologies precisely is difficult as new applications develop with the product, engineering and marketing strategy for a radically new product should emphasize, instead of precise market focus, "functional flexibility." The functionality of the product should be as flexible as can be made by high performance, a wide range of features, and lowest cost. Ryans and Shanklin called this kind of marketing for innovative products "positioning" the product:

> Stated simply, market segmentation is a too narrowly defined term to describe the target marketing activities that need to be employed by the high-tech company. Rather, positioning seems to best describe the steps that the high-tech marketer needs to follow if it is to identify correctly the firms' target markets and to place them in priorities.
> —(Ryans and Shanklin, 1984, p. 29)

Furthermore, E. Peter Ward suggested the following steps of market positioning (Ward, 1981):

1. List and focus on the range of applications possible with the new technology.
2. Describe the present size and structure of corresponding markets for the applications.
3. Judge the optimal balance of performance, features, and costs to position for the markets.
4. Analyze the nature of competition currently in the markets, and strengths and weaknesses of current products compared to the new product.
5. Consider alternate ways in which the new product could satisfy the markets and project product capture of the markets.

6. Consider the modes of distribution and marketing approaches to these different markets that the new product should take.

The notion of market positioning must become almost a kind of experimental approach for really basic new technology. Roland Schmitt has emphasized the need for close cooperation between R&D and marketing when pioneering a new basic technology:

> These [marketing] experts should not, however, give blind allegiance to the latest analytical techniques or the dogma of marketing supremacy. Rather, they should have the temperament of a research experimentalist, putting forth hypotheses about the market and envisioning economical and efficient market experiments.
>
> —(Schmitt, 1985, p 126)

ECONOMIC EVALUATION OF TECHNOLOGY

The economic evaluation of the potential of new technology should be examined using several criteria:

1. Quality—through product superiority
2. Value—through correct market focus
3. Price—through efficient production
4. Opportunity—through timing of innovation
5. Profitability—through investment return

The quality of a product to be delivered depends on its embedded technologies. Will new technology improve the quality of existing products or provide new quality products? The value to a customer depends on the application to which the customer applies the product. Will new technology improve the value or provide new value for customer applications? The price of the product depends not only on the technologies in the product, but also on the technologies in the production of the product. Will new technology reduce the cost of the product?

New technology will provide a competitive advantage to the innovator only when its innovation occurs before a competitor's use of the new technology; thus, timing is important to consider in evaluating new technology. Is there a window of opportunity to gain competitive advantage or, conversely, to catch up with and defend against a competitor's innovation? Finally, new technology must also be evaluated as to the capital required for innovation. All innovations cost money to develop and commercialize. What is the capital required? Can it be recovered in a timely manner? What will the technological innovation contribute to profitability?

TECHNOLOGY DIFFUSION AND SUBSTITUTION

The rate at which a technology enters the market place has been called "technology diffusion." This is the rate at which a new technology is adopted into products, processes, and services. The general form of the rate of technology diffusion is in a S-shaped curve. There are many mathematical models that produce S-shaped curves; for technology diffusion, these have been called *diffusion models.*

If a new technology enters a market by substituting for a prior technology in products, processes, or services, then the new technology is said to substitute for the older one. This has been called "technology substitution." For example, Chris DeBresson tried forecasting the likely first applications of high-temperature superconducting technologies, concluding that medical instrumentation would be the first likely application (DeBresson, 1995).

The rate of technology substitution also follows the shape of an S-curve. Periodically, the literature on technology management provide summaries of the different models of technology diffusion and substition that produce S-shaped curves (Kumar and Kumar, 1992).

The rates of technology progress, technology diffusion, and technology substitution are all s-curve shapes. These shapes can be used for extrapolation forecasts when new technologies begin in the form of Figure 8.1.

The substitution of a new technology for an older technology depends on the new technology providing a higher performance at a quality and cost advantage of the older technology. When the promise of a new technology does not yet attain these advantages, it does not yet substitute. For example,

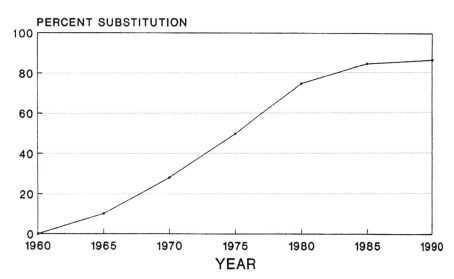

Figure 8.1. S-shape pattern of technology substitution (example is of technology substitution of the float-glass process in Western European countries).

Rebecca Henderson pointed out the unexpectedly long lifetime of the technology of optical lithography in chip fabrication, which, by 1995, still had not been replaced by X-ray printing (Henderson, 1995).

There are many processes by means of which technology substitution and diffusion operate, including purchase of new products, use of vendor expertise, and imitation of other innovating companies' best practices. In particular, technology diffusion of new process technologies proceeds by way of imitation of technology leaders. For example, David McKendrick studied the innovation of worldwide process improvement in banks. He found that banks made use of many channels of information from computer vendors, training, and other sources to innovate new computer-based operating procedures (McKendrick, 1994).

SUMMARY

When technology is changing, there are both technical and commerical risks in innovation. Technical risks arise about factors such as performance, dependability, and safety. Commerical risks arise from uncertainties about who the customers are for the technology, how they will use it, what specifications are necessary for use, how they gain information and access to the product, and what price they will pay.

Financial risks in timing technological innovation arise from the capital required (1) to perform the R&D necessary to design the new product, (2) to produce the new product, (3) to market the new product to the point of break-even on the investment, and (4) to meet competitive challenges as the product and market evolve.

Products embodying technological innovation should contain proprietary technology to provide a competitive advantage. The economic criteria for the evaluation of new technological criteria should use several criteria, including quality, value, price, opportunity, and profitability. The rates of technologies diffusing or substituting into markets show S-shaped patterns.

FOR REFLECTION

Think of an industry and identify one example each of the following kinds of innovation in the industry: radical component, radical systems, incremental, and next-generation technology. Who made them, and what impacts did they have for the innovating firms and on the industry? When did these innovations occur—early or late in the history of the industry?

9

INTEGRATING ORGANIZATIONAL AND TECHNOLOGY CHANGE

CENTRAL CONCEPTS

1. Organizational stasis and change
2. Organization theory
3. Management styles
4. Transilient impact of technological change on business

CASE STUDIES

When Large Firms Became Dinosaurs: IBM, Sears, GM

INTRODUCTION

Although technological innovation can provide a competitive advantage, one of the puzzles about business is why large firms often fail to remain competitive, particullary when technology changes.

We recall that successful large firms usually originate in a new-technology industry by being "first movers," the first to make appropriate investments in advancing technology, large-scale production capacity, national and international distribution capability, and development of management talent to grow the new industry. These first movers are the likely survivors in the core technology life cycle of the industry as it matures. But the conditions for success of the first movers may become exactly the reasons for failure if and when structural changes occur later in the industry. This is how first movers can later get into trouble: Leadership fails to anticipate, prepare for, and exploit new structural changes in the business environment.

We next need to look generally at organizational theory to understand the failure of large organizations to remain technologically competitive. We will briefly review organizational analysis as a tool for helping to manage the impacts of technological change on an organization.

ORGANIZATIONAL STASIS AND CHANGE

It is fundamentally difficult for organizations to deal with change precisely because they are designed to do the opposite—not to change. Organizations are designed to repetitively perform the same functions efficiently and effectively.

> *"Stasis," a steady state of repetition, is the ordinary condition of organizations.*

Organizations are formed and managed for economies of scale—providing many customers and clients with quantities of goods and services. Organizational stasis of a business is aimed at achieving balance between production and demand for products.

Yet, over the long term, when the conditions around a firm change, the organization needs to change. This is the fundamental dilemma of organizational survival: Business organizations designed for stasis must adapt for long-term business survival.

Strategy should determine the goals of a business, and organization the means to the strategic ends. To change goals, leadership needs to formulate a new strategy. To change means, leadership may need to formulate a new organization. A firm should change its organization only when it cannot progress toward its strategic goals efficiently with its current form of organization.

Failure of leadership to anticipate and exploit change is a frequent reason why large organizations get into trouble. Frances Amatucci and John Grant emphasized the importance of strategy focused on long-term organizational renewal:

> The notion that "success breeds failure" is well known to management scholars and practitioners. . . . Today, business enterprises operate in an environment which is increasingly complex and dynamic and the emphasis on strategic renewal has displaced more traditional strategic frameworks. . . .
> —(Amatucci and Grant, 1991, p. 98)

Case Study: When Large Firms Became Dinosaurs: IBM, Sears, GM

This case study illustrates the need over the long term for large businesses to change in order to maintain market dominance. The historical setting of the study is the late twentieth century in the United States, when many major U.S. firms faced new competition from international competitors after their post-World War II dominance of world markets. In the 1980s, the decline in size of three once-dominant U.S. firms—IBM, Sears, and GM—illustrated the need for change. As a popular business journal then put it:

> Dinosaurs? They [IBM, Sears, GM] were a trio of the biggest, most fearsome companies on earth. Here's how earnest executives managed them into historic decline . . .
>
> —(Loomis, 1993, p. 36)

The magazine *Fortune* has listed, in rank order, the 20 largest firms in the world for two decades at 1972, 1982, and 1992. In 1972 and 1982, IBM was in first place, but dropped below 20 and off the list in 1992. This was due to financial problems stemming from technological change. IBM's primary product line, mainframe computers, started to become technologically obsolete in the 1980s. In 1972, Sears Roebuck was in sixth place; it dropped to 13th place in 1982 and off the list of 20 in 1992. This was due to top management's neglect of its core business, retailing. It was surpassed by a new competitor, Wal-Mart—ranking in third place in 1992. General Motors was listed in third place in 1972, dropped to fifth place in 1982, and fell of the list of the top 20 in 1992. This was due to top management's earlier neglect of manufacturing.

In another comparison, IBM's market value in 1992 had fallen below that of Microsoft and Intel, the two firms whose success IBM helped to create: It was IBM's use of Microsoft's MS-DOS operating system and Intel's CPU chip that put these two companies on the road to success. IBM management, however, had underestimated the impact of personal computers.

In the case of Sears, equally poor executive strategy began the company's decline. Sears lost market share in merchandise sales to its competitors, Kmart and Wal-Mart (Loomis, 1993). Sears management did not envision change in the market because its leadership became preoccupied with diversification. Two new competitors (first Kmart, and later Wal-Mart) offered lower prices and took low-price end of the department store market away from Sears. Meanwhile, Sears management continued to distract itself, diversifying into insurance and other financial services. By the 1990s, Wal-Mart had become the most efficient department store operator in controlling inventory and stocking product, using new service technologies in computers and communications.

In the case of General Motors' decline, GM lost substantial U.S. automobile market share to foreign automakers from 1960 to 1992. In the 1960s, GM had deliberately ignored the small-car segment of the market. Top management disliked small cars because they sold for less than large cars, but cost almost as much to make. Accordingly, when GM did produce small cars, it made them spartan and of poor quality. This strategy backfired in the 1970s, when fuel prices soared due to a political cartel among Middle Eastern oil-producing countries. At that time, the small-car market niche in the United States exploded, to grow into the largest portion of the U.S. market. In the late 1970s, GM was caught without desirable models and its market share declined. Foreign producers of autos exported into a growing portion of the U.S. market. By the 1990s, nondomestic automakers held a market share larger than GM's. In addition to disdaining small cars, GM's leadership also neglected modern manufacturing technology and proper organization of production. By 1980, Japanese manufacturers produced automobiles at a faster throughput, with better quality, and at significantly lower cost. GM, along with other U.S. automobile producers, had lost not only automobile leadership in the world, but also manufacturing leadership.

LEADERSHIP AND ORGANIZATIONAL CHANGE

The reason large firms become dinosaurs lies not in the size of brains of the firm, but in how managers misuse their brains—in poor leadership, bad strategy, ignoring change, preoccupation with business diversity, and neglect of core businesses. Firms become dinosaurs when managerial brains fossilize stasis—when they ignore change or fail to react intelligently to change. The economic environment of a firm—business conditions, competitors, technology, market changes—is never simple or static. Changes in technology, competition, and market must be anticipated and proactively challenged; an organization must be altered to meet the new challenges.

Since organizations are designed for stasis—doing the same jobs repeatedly—leadership plays critical roles in organizations at times of change. Proactive leadership is absolutely necessary for organizational change. For example, John P. Kotter observed more than 100 companies trying to change themselves to become improved competitors:

> A few of these corporate change efforts have been very successful. A few have been utter failures. Most fall somewhere in between . . .
>
> —(Kotter, 1995, p. 59)

Kotter saw leadership as the vital role in change, suggesting that several steps were necessary to change an organization:

- Establishing a sense of urgency for change and forming a high-level management coalition for change
- Creating a vision of change and communicating that vision through the organization
- Empowering managers to act on the vision and planning for and creating short-term wins about change
- Consolidating improvements and institutionalizing new approaches for continuing change

These steps all require top-level management leadership. Why leadership in large organizations often fails to perform such steps for change arises from the nature of leadership in large organizations. In large organizations, the leaders usually selected are as those who are committed to doing more of the same. Managers often rise to leadership because they embody a vision of the organization's past, which represents a tested story of success. Past leadership built organizational structures and culture that evolved into a successful company. Later, when the business environment changes, however, the earlier structures and cultures and leaderships become ineffectual in the new conditions.

ORGANIZATION THEORY

To understand how any growing organization evolves into a kind of self-propagating entity (static in structure, culture, and leadership), we need to briefly review the theory of organization.

Modern organization theory represents all organizations as constructed of three aspects: structure, culture, and leadership. *Structure* consists of the explicit forms and procedures of an organization. *Culture* describes the organization's implicit forms and procedures. *Leadership* details the perspectives of people who occupy positions of authority in the organization. In addition to these *organizational aspects,* theorists also have elaborated basic principles about organizations:

1. Vision
2. Division of labor
3. Authority structure
4. Operations structure

5. Organizational rationality
6. Informal structure and conventions

The first principle of organization derives from the organizational aspect of leadership:

1. The essence of leadership in a formal organization is to express a vision of the organization's mission.

Leadership is necessary for any organization because an organization is a goal-directed entity. The goals of an organization are derived from a mission that its productive transformations must pursue. In the case of a business, leadership must express the vision of its enterprise system. To operate within that vision, the enterprise system repeatedly transforms resources (inputs) into products (output). The kinds of repetitive actions are thus determined by the nature of the organization's enterprise.

Thus arises the second principle of organization:

2. Organizations partition repetitive actions as divisions of labor to carry out the enterprise.

Division of labor is meant to increase the efficiency and effectiveness of performing repetitive operations through deepening the skill of the performers and/or automation of the actions. Most business organizations first organize by business function to achieve efficiency through skill, such as divisions of engineering, manufacturing, purchasing, marketing and sales, personnel, finance, and so on. When organizations have more than one product line, they may alternatively organize first by product line and second by function. (For example, in the United States in the 1920s, a new CEO, Alfred Sloan, organized GM into complementary brands in order to cover the price range for automobiles from high to moderate: Cadillac, Buick, Oldsmobile, Pontiac, and Chevrolet.)

To provide the power and accountability to operate the division of labor, organizations also create an authority structure:

3. The authority structure is derived from the need to assign power and accountability in an organization to operate divisions of labor.

Authority structures are hierarchical levels of authority. These levels have the authority and responsibility to plan, decide, and supervise the repetitive activities in the division of labor. An authority structure is typically depicted as an "organizational chart" of the firm. A typical form is shown in Figure 9.1. Organizational charts usually start at the top, with a board of directors, to whom the president of the organization reports. Below reporting to the president, are a set of vice presidents who have the responsibility of op-

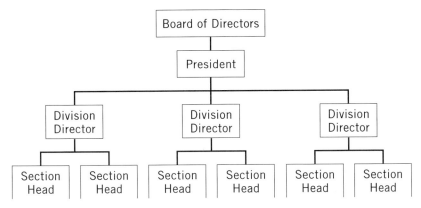

Figure 9.1. Organization chart of an authority structure.

erating the major divisions of the organization. Within each divisional box of the organization chart appear the different functional offices of each division. In principle, the organizational chart descends within the offices to programs and working groups. The individuals who head each organization box of the authority structure are the managers of the organization.

There are two administrative corollaries to this principle of the authority in an organization:

3a. There should be no responsibility assigned to an individual without the accompanying authority to carry out the responsibility.

3b. There should be no authority assigned to an individual without procedures to ensure proper accountability in the exercise of authority.

The fourth principle of organization theory is that, within authority structures, organizations need to develop operating procedures for controlling, performing, and coordinating the repetitive operations into an enterprise system. Together, these procedures constitute the "operations structure" of the organization.

4. The operations structure of an organization consists of an integrated set of policies and procedures that transform organizational inputs into organizational outputs.

Division of labor, authority, and operations together create the "structural aspect" of the organization.

The fifth principle of organization theory is that the structural aspect, combined with leadership, form a kind of "rationality" that transforms organizational inputs into outputs toward the mission of the organization's vision.

5. Organizational "rationality" consists of the degree of efficiency and effectiveness of authority and operations in achieving the mission of the organization.

Efficiency and effectiveness are relative measures, establishing degrees of organizational rationality. Some procedures will be more efficient and/or more effective than other procedures in attaining organizational goals. In management jargon, the procedures known to be the most efficient and/or effective for a class of operations are called "best practice." Best practice is how the most efficient organizations operate a repetitive activity to achieve efficiency in that activity.

For example, Frank Church, in the late 1800s, developed the concept of overhead charges in direct accounting for the cost of production, assigning to each product both direct costs of materials and labor and overhead costs of machinery and factory. This became the "best practice" in industry in the 1900s, until late in the century when overhead costs became larger than direct costs and no longer provided an accurate accounting of the nature of manufacturing costs.

An example of the early 1900s of the principle of operations rationality is the famous approach to management by Frederick Taylor, called "scientific management." Taylor proposed that management should analyze each step in an operations procedure, select the best methods, and determine appropriate times for performance of a step by a best method.

Historically, while Church's and Taylor's approaches to rationality in managing operations were gaining acceptance, another approach to studying organizational rationality was beginning. This was due to sociologists, of whom the most famous is Max Weber. In the early nineteenth century, Weber studied the new government agencies emerging in Europe in the late 1800s. He called these organizations "bureaucracies" and formulated the idea of "bureaucratic rationality" as a kind of operational efficiency.

Weber had compared the new forms of government agencies that grew up in the industrializing Europe to the kinds of governmental offices common in Europe prior to industrialization. Weber called this earlier form a "prebendal" form of office, maintained by "feudal" holders of authority. A feudal officeholder exercised authority in the name of a sovereign ruler in order to perform a governmental function (such as tax collecting, or public order). The nature of the office and the personal property and interests of the feudal officeholder were not separated. The first idea that Weber proposed for a modern bureaucratized office was that the public property and authority of the office should be *separated* from the private property and authority of the officeholder. For example, in the U.S. government, there are federal laws that forbid officeholders from accepting gifts that would create conflict of interest in exercising the responsibilities of the office.

The second idea that Weber proposed for a modern bureaucratized office was that the decision criteria by which decisions are made should be explicitly written down, and the procedures by which activities are conducted should be *formalized*. This explicitness of decisions and formalization of procedures introduced a kind of formal order, "rationality," into the operations of a bureaucracy. Moreover, said Weber, this rational order should be governed by the goal of attaining efficiency and effectiveness in operations. For example, in the National Science Foundation (NSF), a grant-giving agency of the U.S. government, formal procedures require an unbiased and favorable peer review of any research proposal before a research grant may be awarded. Peer review by knowledgeable experts in a research field is held by NSF and its clientele science community as being the most efficient and effective procedure for selecting research proposals for awards.

Influenced by Weber's studies, the *idea* of rules and rationality in large organizations was called "bureaucracy." This was accepted until the 1940s, when other sociologists—notably, Robert Merton—began to study the *inefficiencies* of large organizations. Merton and others gave the idea of "bureaucracy" a bad name. They pointed out that when procedures became formalized, they also became rigid and inflexible. While the formality of procedures promotes efficiency, formalization also promotes rigidity and inflexibility.

One is tempted to ask: Who was right, Weber or Merton? Are bureaucracies inherently rational or irrational? The answer is that both were right. Formalization of decision making and procedures in a large organization does provide rationality and efficiency in operating repetitive activities; at the same time, formalization creates rigidity and inflexibility in policy and decision making. This is the inherent contradiction in formal rationality and in bureaucracy.

The term "bureaucracy" continues to have a bad reputation, although managers in all large organizations, business or governmental, must operate within a bureaucracy. All large organizations require formalization of decision making and procedures. All large organizations must become, to some degree, bureaucratized. The concept of "Weberian rationality" in organizations is the idea of the "benefits" of bureaucracy, while the concept of "Mertonian irrationality" in organizations is the idea of the "inflexibility" of organizations.

The sixth fundamental principle of organization theory is that all organizations develop informal networks of communication and influence and conventions in business practice.

6. Informal structures of groups and conventions within these groups create informal organizational culture and subcultures.

Groups arise in any organization, inherently from the formal structures of the organization. The authority hierarchy and the functional divisions of labor divide the organization into subgroups that can become isolated from one another.

For example, Harold Kerzner presented a symbolic way of showing this organizational division (Kerzner, 1984). Figure 9.2 illustrates Kerzner's idea about group isolation in an organization. In this sketch, there is first a division of the organizational pyramid of authority structure by the grouping of the different levels of authority: the levels of top management, middle management supervisor, and labor. Personnel within each level inherently create group interactions that socially divide an organization into layers, as symbolized by the horizontally sliced organizational pyramid. Also present is a second division of the organizational pyramid, arising from the functional division of the organization, in which different functional subcultures in the organization are created. This is shown by the vertically divided organizational pyramid. Kerzner then suggests that when one superimposes the management gaps of an organization on the functional gaps, one sees how groups of "operational islands" inherently form in an organization. Kerzner then argues that to coordinate such "operational islands" into an efficient organization, systems of procedures and communications are necessary.

In another example, Parry M. Norling discussed informal networks in research organizations, which included informal "content" networks

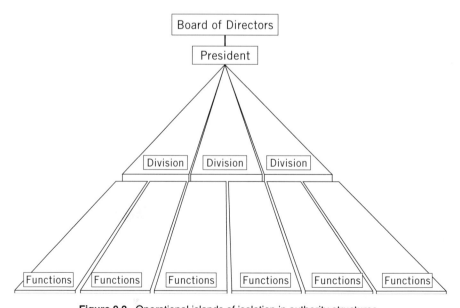

Figure 9.2. Operational islands of isolation in authority structures.

(wherein people communicated about knowledge-based issues) or informal "situational" networks (where people communicated about work conditions or other social issues) (Norling, 1996).

In addition to groups and their processes forming an informal structure complementary to an organization's formal structures, there is also a source of culture arising from the past experiences of managers in an organization. The lessons of these experiences are summed into what Steele called "business conventions" (Steele, 1989). Assumptions underlying the vision, mission, and strategy of an organization complement their formal expression. These assumptions arise partly from the experiences of how an organization became successful. From their experience gained in founding and growing the organization, managers develop a culture of shared beliefs and conventions about how the firm should operate. These conventions include assumptions about the nature of the business, the way competitive advantages are gained, a sense of how and why the company became what it is, procedures for guidance and operational control, and standards of excellence and "how good is good enough." These informal conventions pervade the vision and goals of the organization, the organized division of labor, and the authority and operations structures and procedures.

In addition, to groups and business conventions, there is a third source of informal structure and culture: the impact of the technologies of production on management priorities. Tom Burns and G. M. Stalker pointed out that the culture of an organization is partly dependent on the nature of technical operations. They proposed two extreme types of organizational cultures, mechanical and organic, depending on the nature of the operational control issues (Burns and Stalker, 1961).

When an organization produces large volumes of one product for many customers, then production processes are standardized and the primary management problem is exact control of repetitive operations. According to Burns and Stalker, this concern for exact control over similar and repetitive operations creates a "mechanistic" management culture. Managers share the convention that the organization is like a complicated machine that requires continual attention to details to run efficiently. Examples of industries in which "mechanistic" management cultures arise are large-volume producers, such as automobile, computer, and appliance manufacturers.

In contrast, for an organization that produces a one-of-a-kind product for a customer, production operations are not repetitive but tailored to each specific project. In this case, managers are concerned mostly about delivering a customized product on time and within budget to satisfy a specific customer. According to Burns and Stalker, this creates an "organic" management culture. Managers share the convention that the organization should be focused on specific and finite projects, with primary attention to each client. Examples of industries in which organic management cultures can arise are client-oriented firms such as advertising firms, law firms, and hospitals.

In summary, organizational theory presents organizations as complex sociotechnical systems with both formal structure and informal cultures. Organizations can behave irrationally as well as rationally and can be economically efficient at one time but become economically inefficient at a later time.

MANAGEMENT STYLES

In addition to the principles of organization, we also recognize three styles of management: bureaucratic, entrepreneurial, and project management.

Bureaucratic style (in the "good" sense of "bureaucracy," as "rationality") is the appropriate management style for running large organizations. Large organizations require rational formalization of the organization to effectively and efficiently focus large-scale, repetitious activities on the mission of the organization. The problem bureaucratic style faces is to avoid rigidity and empire building (which gives bureaucracy a "bad" connotation). All large organizations get established as bureaucracies (even the "flatter, leaner, meaner" kind of organization in vogue in the early 1990s).

However, starting up a new organization requires a different management style, which is focused on building an organization; this has been called an "entrepreneurial" style. This style is essential to the launching of new high-tech business ventures, and we will review it in a later chapter.

Distinct from these two styles is a third style for introducing finite change in an existing organization; this has been called a "project management" style. Project management is essential to implementing innovation in research and development projects and in design and service projects, and we also will review this style in another chapter as well.

TRANSILIENCE OF INNOVATION ON BUSINESSES

Organizational structure, culture, and leadership can all be affected by technological changes. William Abernathy and Kim Clark emphasized the broad range of impact a technological change can have on a business—which they called the "transilient" impact of technology (Abernathy and Clark, 1985). Change in the value-addedness of new technology can pass through the entire business system—changing not only the kinds of products and the way they are produced, but also the kinds of customers and markets the business serves.

Furthermore, Abernathy and Clark classified the transilient aspects of innovation by their impacts on the core competencies of the firm—either product/production or market/customer competencies:

1. In product/production competency, innovations may alter:
   a. Product design
   b. Production systems
   c. Technical skills and knowledge base
   d. Materials and capital equipment

2. Under market/customer competency, innovations may alter:
   a. Customer bases
   b. Customer applications
   c. Channels of distribution and service
   d. Customer knowledge and modes of communication

The impact of innovation on any of these factors may range from strengthening existing competencies to making existing competencies obsolete. Accordingly, Abernathy and Clark classified technological innovations as to how they conserved core business competencies or rendered them obsolete:

1. A technological innovation that conserved both existing production and market competencies was called a *regular innovation.*
2. A technological innovation that conserved existing production competency but altered market competency was called a *niche-creation innovation.*
3. A technological innovation that made existing production competency obsolete but preserved existing market competency was called a *revolutionary innovation.*
4. A technological innovation that made both existing production and market competencies obsolete was called an *architectural innovation.*

Historically, firms have usually successfully exploited regular or niche-creation innovations. But many large firms have perished during revolutionary or architectural innovations in their industry.

For example, as we saw in a previous case study, IBM met with an architectural innovation in the new technology of the "distributed computational system," which resulted from hooking personal computers together in communications networks. The distributed computational system technologies made both IBM's existing production competencies in mainframes and its existing market competencies in centralized information system offices obsolete. Sears met up with a niche-creation innovation, in which the new service technologies of computerized point-of-sales and ordering enabled Wal-Mart not only to enter the market niche of country-based retail, but also to move into suburban-based retail with a cost advantage over Sears. GM also met with a niche-creation innovation in the small car market, and encountered revolutionary innovation in the new manufacturing techniques innovated by Toyota.

These examples illustrate the importance of management anticipating and altering strategy and organization when their business environment changes due to transilient innovations.

REDESIGNING ORGANIZATIONAL PROCESSES AND STRUCTURES

Technological discontinuities that render production and marketing core competencies obsolete will require a change in business strategy at the CEO level and implementation of reorganization to effectively carry out such a strategy. To integrate organizational change with technological change, one can use the following logical steps:

1. Organizational change should be anticipated as necessary when an affecting technological change is "niche-creating," "revolutionary," or "architectural."
2. This level of change will require a CEO-level vision and commitment to implement change in the organization.
3. Long-term product-line planning will be required to create new or refined products to exploit the technology change.
4. Long-term marketing planning will be needed to create new or refocused markets for the new or refined product lines.
5. Long-term manufacturing planning will be necessary to produce new or refined products with a focus on cost and quality for eventual commodity-type products
6. Organizational structure and procedures will need to be redesigned to facilitate changes in technology, product, production, and market changes.
7. Information systems will need to be redesigned to facilitate communications, decisions, and coordination in new organizational structures and environments.
8. Personnel education, training, hiring, and reward measures will need to be redesigned to facilitate new technology and transilient effects.
9. Financial plans will need to be reformulated to finance and control appropriate returns on changes.
10. Stakeholders must be appropriately prepared for the long-term and extensive nature of transilient changes.

Strategic change in organizational structures can occur only periodically for organizations to operate effectively. Strategic organizational change temporarily disrupts operations. For example, Marcie J. Tyre and Wanda J. Orlikowski studied the pattern of process change in organizations and concluded that changes occur in groups, with periods of no change between them (Tyre and Orlikowsi, 1993).

SUMMARY

It is difficult for large firms to deal with structural change precisely because organizations are designed to do the opposite—to repetitively perform similar functions with efficiency. Failures of leadership to anticipate and exploit change are why large organizations set up for large-volume repetitive operations eventually get into trouble.

Modern organization theory represents all organizations as constructed of three aspects: structure, culture, and leadership. These conditions create stasis, which is successful for the conditions in which the organization originally evolved. Conditions for success of industrial first movers become the reasons for failure when later structural changes occur in the industry.

We now recognize three different styles of management: bureaucratic for managing stasis, entrepreneurial for managing new business startup, and project management for managing finite change.

Technological change can have a transilient impact on a business when its impacts pass through the enterprise system—possibly changing not only the kinds of products and the way they are produced, but also the kinds of customers and markets the business serves. When technological change affects any of the subsystems of the enterprise system, then redesign of operations of affected subsystems must to occur.

FOR REFLECTION

Find an historical account of a large firm no longer existent, but in which technological change was important to its origin and death. Describe an organizational life cycle. What roles did business and technology strategies play in its origin, growth, temporary market dominance, and eventual demise?

___10
MANAGING WITH TECHNOLOGY

CENTRAL CONCEPTS

1. Technologies and management challenges
2. Operations structure
3. Software tools for operations
4. Authority structures in organizations
5. Software tools for authority
6. Redesigning organizational processes
7. Influence of technology on management paradigms
8. Technology-based management practice

CASE STUDIES

Five Companies Benefit from Information Technology in the 1990s
Influence of Technology on Management Paradigms

INTRODUCTION

We have reviewed organizational theory to understand how technology becomes embedded in organizing business processes. Now, we need to look in more detail at the kind of service technologies that have emerged for managing organizational processes. We also need to understand how, in general, changes in technology can affect the practice of management itself.

Historically, technological change has always had an impact on management practice. Management practice is about how to plan, capitalize, organize, operate, and control a business based on productive technologies. As productive technologies changed, new management challenges were created that stimulated new ideas about management—changes in the paradigm of management practice. A *management paradigm* is a set of concepts about what constitutes good management practice. For managing technological innovation, it is important to appreciate how new technical opportunities may require changes in management practice.

TECHNOLOGIES AND MANAGEMENT CHALLENGES

We recall that the progress of technologies over the last two centuries has been in both technologies of "power" and of "thinking." The newer technologies of thinking, centered around the computer and communications, are altering the economies of the world and fostering new management paradigms. Previously, the concern with powered production focused on managing technology—managing technological innovation for new products, processes, and services. Now, the concern for computerized production is adding a focus on managing with technological innovation in computerized design, production, communication, and information technologies. Ritchie Herink has called this newer change in management practice: "managing with technology" (Herink, 1994).

In the 1980s and 1990s, digital and distributed computational and communications systems for management information and control began altering the hierarchical structuring of large organizations—leading to flatter structures and eliminating some layers of middle management. In engineering and manufacturing, computers and communications began leading to virtual product development techniques and computer-integrated and agile manufacturing techniques.

Also, at this time more systematic attention was paid to the pervasive effects of rapid technological change in a mixed set of core technologies in products, processes, and services. For example, machine tools have traditionally depended on two critical core technologies, metallurgy and mechanical machinery. With the advent of numerically controlled machinery, however, a third critical technology has been added—electronic control. The rate of change of technology in metallurgy and in mechanical machinery has been relatively slow and incremental, but the rate of change of technology in electronic control has been fast and discontinuous. Therefore, the pacing technology for the recent generations of machines has been in electronic control.

Case Study: Five Companies Benefit from Information Technology in the 1990s

This case study illustrates how investment in information technologies began to improve the operations of firms. The historical setting of the study is the early 1990s, after the technical changes of the 1980s which created the first distributed computational systems, and these began to have dramatic impacts on organizational structures. By 1993, capital investment in information technology was the fastest growing category of capital expenditures in industry. As a popular business journal then reported:

> Though barely out of its infancy, information technology is already one of the most effective ways ever devised to squander corporate assets. Year after year, the typical large business invests as much as 8 percent of revenues in telecommunications, computer hardware, software, and related high-tech gear. Information technology soaks up a dramatically growing share of corporate spending, accounting for over 14 percent of existing U.S. capital investment as of last year [1992] versus 8 percent in 1980.
> —(*Fortune*, 1993, p. 15)

The management problem with information technology investments was: Where was the improvement in productivity? Traditionally, capital investments in production facilities either increased production or reduced the amount of labor input per unit product ("productivity"). But did information technology always increase production capacity or improve productivity? The answer was—not always. For information technology could also provide other economic benefits, such as improved quality, reduced overhead, faster organizational response, improved organizational control, and expanded product lines. Just how a given investment in information technology provided economic benefits depended on how the technology was used.

For example, in 1993, *Fortune* magazine studied five companies that then had reaped economic benefits from their recent information technology investments:

> We identified five—Ford Motor, Dell Computer, Caterpillar, Chris-Craft, and American Express—that have successfully applied information technologies to familiar business problems.
> —(*Fortune*, 1993, p. 15)

At Ford, information technology investments had produced the capability of integrating design activities throughout its global structure. Ford integrated computer-aided design tools to connect several design studios that were located around the world. In new designs of vehicles Ford was achieving:

> [S]ubstantial cost savings and smarter allocation of scarce resources, both human and mechanical. . . . These computers transform designer's sketched lines into mathematical models of a vehicle's entire surface. . . . The computers . . . can produce the numerical codes to control the milling machines that carve styrofoam or clay into full-size mockups of cars. Plugged into the automaker's $35-million-a-year global telecommunications network, the system allows designers in seven studios all over the globe to work together. . . .
>
> —(*Fortune,* 1993, p. 16)

Dell's application of new information technology was for marketing. Employees receiving "800" calls from customers about purchases were linked by network to a parallel computer, which archived information about a customer in a single database that was accessible by all employees. Dell personnel used the database to increase sales and customer satisfaction:

> For example, the rate of response to its small-business mailings rose 25 percent once Dell customer feedback was used to refine its pitch. Experience from the database guides the sales representatives who receive calls. Product developers rely on the database to help them shape new offerings. . . . Routine analysis of sales information helps Dell spot such consumer trends as the shift to larger hard disk drives.
>
> —(*Fortune,* 1993, p. 22)

In another example, Caterpillar used information technology to improve its inventory control:

> Caterpillar, which returned to profitability this year [1993] after seven quarters of losses, is typical of old-line manufacturers that reengineered key processes in response to reverses in the marketplace. Gored in the mid-1980s by Japanese competition, rising manufacturing costs, and declining market share, Caterpillar launched a $1.85 billion program to modernize its 17 factories. The goal was to increase product quality and plant flexibility while slashing inventories and production-cycle time. The company recouped its costs last year, and since then the return on its investment has been running at a 20 percent annual rate.
>
> —(*Fortune,* 1993, p. 22)

In yet another example, Chris-Craft used information technology to improve product design. In 1988, U.S. pleasure boat market sales peaked, and Chris-Craft was plunged into bankruptcy:

That vortex sucked into bankruptcy one of the world's best-known boat-makers, once renowned for the sleek mahogany speedboats it sold to movie stars. Chris-Craft emerged chastened, with a new corporate parent, Outboard Marine of Waukegan, Illinois, and a new plan for survival . . . with a strategy to introduce new products to gain share in a market that is now essentially flat. To do that, Chris-Craft laid out $300,000 for a computer-aided design system. . . . The primary payoff from CAD is a dramatic cut in product-development cycles and lower manufacturing costs.

—(*Fortune,* 1993, p. 24)

In the fifth example, in 1991, American Express was spending more than $1 billion annually on information technology to improve operations in its credit card business. It then began an experiment to allow its most reliable agents to work from their homes. It proved so unexpectedly successful that American Express plans to expand it, with, eventually, 10 percent of the 10,000 American Express employees who do telephone order-entry work using it. It cost American Express $1300 as a one-time expense to connect home-situated employees to its phone and data lines. Then, calls bounced seamlessly from its reservation center to its agents' homes, where the agents look up fares and book reservations on their PCs. Supervisors can monitor the agents' calls, ensuring control:

The productivity gains so far have been astounding. The typical agent— they're all women so far—handles 265 more calls at home than at the office, resulting in a 47 percent average increase in revenue from travel bookings, or roughly $30,000 annually each.

—(*Fortune,* 1993, p. 28)

OPERATIONS STRUCTURES

We recall that an operations structure is one of the principles of organizational analysis. We will now use the concept to identify opportunities for information technology to contribute to the management of the organization—managing with technology. Operations structures can be modeled as three kinds of operational flows in an organization:

1. Flow of direct value-adding activities
2. Flow of scheduling and control information
3. Flow of indirect value-adding activities

As illustrated in Figure 10.1, these three operational flows can be symbolically depicted as three simultaneous flow diagrams on parallel planes:

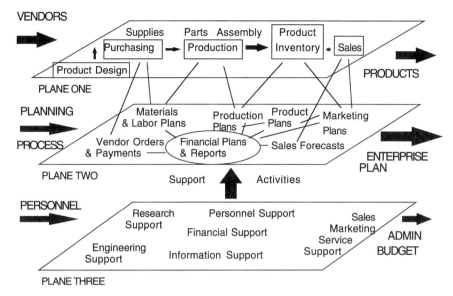

Figure 10.1. Operations structures.

1. Production transformation plane
2. Financial control plane
3. Support activities plane

Plane One constitutes the flow of activities for the production operations of an organization, representing its direct value-adding activities:

1. Purchasing buys parts, materials, and energy for production.
2. Materials and parts are received into the materials inventory or directly to the production line.
3. Production assembles products.
4. Work-in-progress inventory holds partially completed products.
5. Completed products accumulate in product inventory.
6. Sales of product draw on products in inventory.
7. Sold products are distributed to wholesalers, retailers, or customers.

Plane Two constitutes the flow of financially relevant information that schedules and controls operations:

1. Sales and sales forecasts receive and anticipate income and determine production plans.
2. Marketing plans create strategies for sales.

3. Production plans schedule production and determine materials and labor requirements.
4. Materials and labor plans determine supplies acquisitions and schedule labor.
5. Vendor orders and payments place materials orders and schedule payments.
6. Financial plans and reports plan profits and working capital and report costs, sales, profits, and balances.

Plane Three constitutes the flow of the overhead activities that support the production plane of the firm:

1. Research support
2. Engineering support
3. Sales, marketing, and service support
4. Personnel support
5. Financial support
6. Information systems support

The lines between the three planes indicate information connections.

MANAGEMENT TOOLS IN OPERATIONS STRUCTURES

With this scheme of the operational flows within a productive firm, one can identify where information tools for management can be used:

1. Plane One
   • Materials resource planning software
   • Production scheduling software
   • Computer-aided manufacturing software
   • Inventory control and distribution software
   • Point-of-sales software
2. Plane Two
   • Accounting systems software
   • Planning and budgeting software
   • Executive information and control software
   • Vendor electronic ordering software
   • Investment management software
3. Plane Three
   • Project management software
   • Computer-aided design and engineering systems software

- Personnel records and evaluation software
- Training software
- Marketing analysis and forecasting software

Thus, one can see the range of software tools that have been developed for assisting modern management of operations in a productive organization. For example, James Brian Quinn, Jordan J. Baruch, and Karen Anne Zien have emphasized the importance of using software innovation for improved organizational processes (Quinn, et al., 1996).

MANAGEMENT TOOLS IN AUTHORITY STRUCTURES

We recall that in addition to operations structures, all organizations have authority structures. Authority structures order the relationships of power between divisions of the organization by defining scopes and levels of responsibility, accountability, and authority. We can use this concept to further identify where information technology assists management to manage with technology. Management software tools for authority structures are aimed at facilitating planning, coordination, and accountability functions. Examples include the following:

Planning tools:
- Relational database software
- Spreadsheet software

Coordination tools:
- Project scheduling software
- Electronic mail software
- Phone systems software
- Teleconferencing systems software
- Groupware

Accountability tools:
- Accounting systems software
- Metrics monitoring systems software
- System activity monitoring systems software
- CEO database and monitoring systems software

RE-ENGINEERING BUSINESS PRACTICES

Improvement of business practices through deliberate re-examination and improvement of specific procedures will continue to be important to improving the technology of business operations. The continuing pace of computational and communications technologies means that, periodically, businesses need to look at their operational and authority structures and re-engineer business practices for improved efficiency and effectiveness. D. Jack Elzinga, Thomas Horak, Chung-Yee Lee, and Charles Bruner have described the logical steps for improving specific business procedures, which include process selection, description, quanification of process parameters, improvement selection, and improvement implementation (Elzinga et al., 1995)

The concept of re-engineering business practices developed a bad reputation in the early 1990s in the United States, when it was used as a corporate excuse for downsizing—firing employees. Redesigning business processes to take advantage of new service technology capability need not result in firing employees; this will occur only under poor management.

In fact, the history of automation, both in manufacturing and in services, has shown increased employment when automation improved business competitiveness. What changed was not the total number of employees, but the balance of the kind of employees—always with an increasing demand for employees with higher job skills and training.

What technological progress generates, both in power and thinking technologies, is an increasing professionalization of the workforce.

Case Study: Influence of Technology on Management Paradigms

It is important to take a broad picture of how management practice alters as a result of technological change. The change in what is thought at a time to be the best management practice has been called a "paradigm" of management. In this case study, we will briefly review the history of change in paradigms of management to illustrate how, in fact, concepts of management altered to meet operational challenges that new technology opportunities created. The historical context is the evolution of management thought over the past three hundred years of the industrialization of the modern world.

The modern concepts of what constitutes an enterprise and how to manage an enterprise date from the first industrial revolution, which began in England in the late eighteenth century. Before that industrial revolution, economic production was organized in Europe as guilds located in cities, and in the form of economically self-sustaining manorial systems. After the industrial revolution, economic production began to be organized into factory systems located in cities.

We recall that the key technological change that made the factory system imperative was the innovation of steam-powered production machinery. From a management perspective, this technological change in production made it obvious that the most valuable production asset at the time was the machinery. With the importance of this value-adding asset in mind, management turned from hiring labor on a piece-rate basis to a time-employed basis. Shifting to the time-employed basis meant that management would have to both prescribe and supervise how labor's time was spent.

As the new production technologies made mass production possible, the next challenge for management was how to control quality in production. The management idea of "interchangeable parts" was pioneered by Whitney Hays in U.S. product production. This provided a basis for another shift in management paradigm. By the end of the nineteenth century, accounting for the costs of production and measuring profitability became important. So a third major shift in the paradigm of good management practice occurred when Frank Church innovated the accounting practice for the costs of production to be based on direct labor and materials, and all else as overhead proportioned to the direct costs. The technologies of production required changes in the ideas about industrial organization. Technology of the production operations also altered the ideas about costing operations.

The continuing problem about how to prescribe and supervise time-based labor as increasingly complicated production technologies were invented stimulated a fourth major paradigm shift by the end of that century, with Frederick Taylor's concept of "scientific management." Taylor argued that all productive operations could be analyzed and optimized as unit actions and sequences of unit actions. Again, the increasing complexity of the technologies of production challenged management ideas into of conceiving of production management as a scientifically-based set of industrial procedures, and gave rise to the engineering discipline of industrial engineering.

One unfortunate consequence of Taylor's conception of the appropriate roles of management and labor, however, was to lose the contribution that labor can and should make to decisions on productive operations. After Taylor, a too-narrow concept of scientific management became popular and contributed to poor management-labor relations in Europe and in America. Managers began to think of themselves as the "brain" of the production organization, and of labor as the "brawn." This led to serious problems with quality and quantity of production, by making management-labor relations antagonistic and confrontational.

Even much later, after Elton Mayo's studies clearly demonstrated the inevitability of labor participation in decisions on production, the idea that worker cooperation was essential to production efficiency and quality was generally ignored by American and European management until long after the Second World War. Only late in the twentieth century did Japanese management demonstrate to the world the importance of including labor in production decisions, with a view toward motivating workers to contribute to the continuous improvement of production quality.

But long before this refinement of management thought occurred, a fifth major paradigm shift occurred in the 1920s in the United States when

Henry Ford introduced the concept of the assembly-line organization of production.

PRE-WORLD WAR II MANAGEMENT PARADIGM

Up to the Second World War, the dominant management paradigm about production emphasized the following principles:

1. Reducing the direct costs of production should be the primary focus of management concern.
2. Management should be regarded as decision makers and labor as passive followers of instructions.
3. The operations of an enterprise could be analyzed as a stable set of unit operations.
4. Production economies require large volumes of standardized products on assembly lines with fixed automation.
5. Single critical-technology-based product lines will have long product lifetimes, providing long periods of stability in the organization.
6. World markets can be divided on a national basis, with national firms dominant in domestic markets.

Case Study (Continued): Changing Management Paradigms of the Later Twentieth Century

The evolution of Japanese management practice, as we indicated, provided a sixth major paradigm shift in management thought. For example, Michael A. Cusumano summarized several management innovations made by Japanese management in the automotive industry: just-in-time manufacturing, reduction of process complexity, and vertical de-integration with stronger supplier relationships (Cusumano, 1988). As another example, Walter Robb of General Electric Corporation summarized the Japanese management contribution to the product development process, which GE learned about:

> [P]roduct development planning was a team endeavor.... They weren't aiming for a product with the best possible features, but one with the low cost that would satisfy the customer's needs and be easy for the customer to use.
>
> —(Robb, 1992, p. 8)

The emphasis on quality improvement through teams and innovation can lead to great advances over time. As production technologies continued to increase in technical complexity, the knowledge gained from working with the details of the production technology became increasingly important to gaining competitive production quality.

However, as production technologies matured and competition created excess production capacity, another strategic issue became evident—how to compete in producing for a crowded marketplace. So, a seventh paradigm shift occurred, from Toyota management's leadership in providing strategy about improving profitability in low volumes of production. Toyota innovated new management techniques for flexible production, just-in-time production, and rapid production response.

Next, the innovation of computer technology and its application to production control in the second half of the twentieth century also stimulated new management thought about industrial organization and production. We call this eighth change in management paradigm "managing with technology." Computer-based service technologies began providing both internal services and external service technologies for management control. Internally in a firm, computers have provided new service technologies for accounting, inventory control, transportation, distribution, production scheduling, retail scheduling, analysis, and engineering design. Externally to a firm, computers have provided service technologies for transactions, scheduling, and communications.

Finally, as the twentieth century closed, the economic benefits of technological innovation could be captured nationally as the new technology diffused more slowly throughout the world. After the Second World War, national policies turned to the deliberate borrowing and developing of critical technologies on a national scale. The United States focused on defense technologies and funded them with large-scale federal investments in R&D. Japan and Germany reconstructed their economies with national efforts focused on civilian technologies. Companies in these nations aggressively borrowed new technologies from other countries, and improved and implemented these in new industries. Thus, a tenth shift in management paradigm occurred. As a result, the pace of technology diffusion in the world increased dramatically, resulting in a globalization of technology, production, and markets (which we have discussed previously).

TECHNOLOGY AND MANAGEMENT PRACTICE

From the history of changing management paradigms, one can see that it is important to include technology within a paradigm of management. Neglecting technology means not paying proper attention to the value-adding knowledge base of the business. This is a grievous error, but one that is easy to make when managers get preoccupied solely with the shorter-term concerns of finances and sales. Technological change is the key ingredient in the strategic management of the enterprise, managing its change.

A management paradigm, which includes the ideas of both managing with technology and managing technological factors of the enterprise, should formulate enterprise strategy that focuses on both kinds of value-adding properties of technology—production and management. This paradigm should attend to:

1. Reducing the indirect costs of the enterprise as well as the direct costs
2. Decentralizing multifunctional and multidisciplinary enterprise teams to decide and operate the productive activities of the enterprise
3. Ensuring that the operations of an enterprise are flexible, agile, and continuously improvable in successive states of quasi-stable production conditions
4. Making production economies of scope equally important with economies of scale and production automation appropriately balanced between hard and soft automation (depending on product volumes and lifetimes)
5. Knowing that multi-core-technology product lines will have shorter product lifetimes and should be planned as generations of products (paced by the most rapidly changing critical technology), and the organization must be flexibly organized to rapid and correct response
6. Realizing that world markets and technology are now global and enterprises should be globally based to "think globally and act locally"

Accordingly, the following principles provide the focus for a new paradigm for management, which includes the lessons from the past and adapts them to modern conditions. They incorporate managing not only the manufacturing, marketing, finance, and personnel factors in the business enterprise, but also technology. An enterprise must be:

1. Value-creating
2. Responsive
3. Agile
4. Innovative
5. Integrating
6. Professional
7. Fair

The basic social responsibility of the enterprise is to provide *value-creating* activities for society. The difference between legal and illegal enterprises is the concept that the products produced for society add value to society and do not subtract value. (For example, the failure in the twentieth century of the Communist approach to economy was its inability to foster value-creating activities in the economy for all except the ruling few.) The focus should be on the nature of the enterprise's products (goods, processes, or services) and how they add value for the customer; this is the key to the long-term survivability of the enterprise.

Responsiveness is necessary to the enterprise in a world of fast changing conditions. A modern enterprise must not only be able to manage for stability, but must also manage for change. To do this, the enterprise needs short cycles in all of its operations, from product development and production to distribution and marketing.

To gain profitability from quick response to changing opportunities both in technology and in the market, *agility* in production capability is necessary. Computer-based technologies in manufacturing are making possible (1) production facilities that are flexible to produce a variety of product lines and (2) utilization of suppliers' production facilities to produce the variety of parts required for agile production.

Since all technological capability eventually diffuses and all new technologies mature, no firm can continue to gain a technological competitive advantage without the ability to innovate. *Innovation* produces the distinct competencies that provide superior products, production, and services over competitors.

Most modern products, production, and services require several technologies as core to their design and production. A modern firm must be able to acquire and *integrate* technologies using different skill bases and knowledge in engineering and science. Without the capability of integrating technologies, a modern firm cannot dominate competition in products having rapid product cycles.

Both economic and technological activities are based on knowledge and experience. Therefore, the intellectual quality and commitment of the workforce in the enterprise will always determine the difference between competitive and noncompetitive enterprises. A competitive firm needs to hire and encourage a workforce with high levels of training and continual upgrading of skills, and with a professional attitude toward work and teamwork—a *professionalized* workforce. The old antagonisms between management and labor are too inefficient and costly for a modern firm to be competitive. The flatter nature of organizations, the higher levels of skills required for production and marketing, and the need for cross-functional teaming for decisions and operations all require managers develop a professionalized workforce.

Finally, the firm surviving over the long term must now create conditions of *fairness* in the type and distribution of the wealth created by the success of productive operations. The "stakeholder" theory of the firm as a successor to the older paradigm of owner/operator and labor is no longer a nicety, but a necessity. Executive teams, who have excessively controlled corporate boards to provide themselves with very large and disproportionate rewards from profits, have consistently, over the long term, been shown to have serious problems with (1) maintaining a cooperative and efficient workforce, (2) maintaining share prices to prevent hostile takeovers, and (3) maintaining public credibility to prevent punitive government regulation.

SUMMARY

Technological change has always had an impact on management practice. Both technologies of power and of thinking have had great impact on stimulating new management thought.

For example, software technologies for operations structures can be used to improve the flows of direct and indirect value-adding activities and for schedualing and controlling these activities. Other software technologies can be used for managing authority structures in planning, coordination, and evaluation. The continuing pace of computational and communications technologies means that, periodically, businesses need to look at their operational and authority structures and re-engineer business practices for improved efficiency and effectiveness.

One needs to take a broad picture of how management practice is altered as a result of technological change, since new technology opportunities often create new operational challenges. A modern management paradigm now needs to attend to value creation, organizational responsiveness, production agility, technology innovation, technology integration, professionalization of the workforce, and fairness of distribution of corporate benefits.

FOR REFLECTION

For a firm or division for which you work or are familiar, diagram the operations structure of the unit. Diagram the authority structure. Does that firm or unit have available a diagram of the operations structure? Of the authority structure? If both are not available, why not? What kinds of management software are currently used for either structure? What new kinds could be used?

───11
TECHNOLOGY FORECASTING AND PLANNING

CENTRAL CONCEPTS

1. Rates of change of technologies
2. Changing the natural phenomena underlying a technology
3. Changing the inventive logic of a technology
4. Extrapolating a technology S-curve
5. Anticipating the natural limits to a technology
6. Proceeding from forecasting to innovation
7. Morphological analysis

CASE STUDIES

Inventing the Germanium Point-Contact Transistor
Inventing the Silicon Transistor
Inventing the Integrated Circuit Chip
Progress in Semiconducting Chips
Progress in Vacuum Technologies
Xerox Corporate Research Organization in 1981

INTRODUCTION

Let us now return to the general concept of technology, particularly how the rate and directions of technological progress can occur. We have emphasized that, in technological innovation, timing is a critical factor in gaining commercial benefits. There is a "window of opportunity" for an innovator to gain competitive advantage and commercial benefit. Accordingly, it is important to anticipate what the technological change might be and when it might occur.

Techniques to anticipate technological change have been called "technology forecasting." Actually, this name may be misleading, since technological change is not something natural (e.g., as the weather, which changes by reasons of physical processes). Technological change results from deliberate human activity—anticipating, planning, and implementing. The first step in creating technology change is anticipation, and the second step is technology planning. Therefore, by the term "technology forecasting," we will mean "technology anticipation."

TECHNOLOGY FEASIBILITY

Technology forecasting is not science fiction. Merely envisioning a new technology is not the same as forecasting it. For example, as we recall, although Leonardo da Vinci envisioned the helicopter, he did not invent it; it was Sikorsky who invented the helicopter. Thus, forecasting cannot be simply vision, science fiction or otherwise. It is true that good science fiction writers have had extraordinary vision. For example, Jules Verne, in the nineteenth century, envisioned many later inventions, such as the submarine and the trip to the moon. But he did not invent anything. Moreover, Verne's envisioned "time machine" has not yet been invented (and will never be invented until we discover something about nature that makes it scientifically feasible). This must be the beginning of any technology forecasting—knowledge that such technological change is scientifically feasible.

Case Study: Inventing the Germanium Point-Contact Transistor

To be innovative, a manager must anticipate technological change. This case study illustrates how an innovative R&D manager fostered a major technological invention that revolutionized electronics in the second half of the twentieth century. The illustration also shows the deep connections between new science and radically new basic inventions. The historical setting is in the 1930s at an industrial laboratory, Bell Labs of American Telephone and Telegraph (AT&T).

Electronics was the new technology in the beginning of the twentieth century, made possible by a key technological component, the electron vacuum tube. The science underlying the vacuum tube had been discovered in the 1800s, when physicists saw that electrons could travel in a vacuum. They constructed a glass tube, inserted metal plates at each end of the tube, evacuated the air from the tube, and then, by applying a voltage across the two tubes, observed an electrical current through the tube: A stream of electrons boiled off the negatively charged plate of the tube and accelerated toward the positive plate of the tube. An inventor, Lee De Forest, added a small grid of wires perpendicularly midway between the negative and positive plates, and found that a small voltage between the grid and negative plate had a huge influence in the size of the current between the negative and positive plates. This small grid voltage controlled the larger plate current. It made the electron vacuum tube into a type of "control valve" for controlling the flow of electrical current and the attendant voltage drop across the plates. The electron tube became a "voltage control valve." The first use of voltage control was to amplify electrical voltage signals; this allowed the development of long-distance telephones, radio communications, and so on.

In the United States in the early twentieth century, AT&T had become the centralized U.S. telephone company and needed to provide long-distance telephone service. Engineers and scientists at Bell Labs leapt upon the electronic tube and innovated the first mass production of electron vacuum tubes for amplification of phone conversations. Later it was apparent that there would be natural limits to the new artifact, for electron tubes consumed a great deal of energy to heat the negative plate to boil off electrons, and electron tubes were large, were not very rugged, were sensitive to overheating, and did not last long, because residual gases in the tubes eventually assisted in the corrosion of the wires in the tubes. By 1935, some of Bell Lab's scientists were already anticipating replacing the electronic vacuum tube with a better device, but what would it be?

Nature provided a clue. One of the early inventions in another new technology was the discovery that germanium crystals could be used to detect radio signals. Moreover, it was found that silicon crystals could convert light into electricity and change alternating current to direct current. These interesting materials were called "semiconducting" materials. Scientists at Bell Laboratories set up a research program, "solid-state physics." The reason for this name was that a new theoretical advance in physics in the 1930s—quantum mechanics—was providing a new way to understand electronic properties of solid materials (Bardeen, 1984).

World War II interrupted the research, with Bell Lab scientists turning to research on radar for the United States. After the war, the solid-state physics group was reconstituted, with William Shockley as head. John Bardeen and Walter Brattain were also members of the group. Shockley's group learned by that doping (adding) other atoms into the germanium crystal during the crystal's growth, electricity would be conducted by the extra electrons (negative carriers) introduced by the other atoms (or by the deficit of electrons, positive carriers, introduced by different atoms). Semiconducting materials are naturally nonconducting, or "insulators;" the other introduced atoms made these materials conducting. So, by "doping" (deliberately introducing other kinds of atoms) into germanium or silicon, one could control the type and amount of conductivity—hence, the name "semiconducting."

In the summer of 1945, Shockley proposed controlling the current in doped germanium crystals by applying an external electric field perpendicular to the direction of current in the crystal. This was an analogy to an electron vacuum tube, trying to control the current through the semiconducting material to make another kind of "control valve." The reason Shockley thought this was practical was that he had used the new quantum mechanics theory to calculate the expected effect of the controlling field on the electron flow. His calculations showed that the effect would be large enough to provide amplification. Amplification meant that the electron flow through germanium would closely follow the increases and decreases of the controlling field, thereby mimicking its form.

Shockley constructed a transistor, but it didn't work—it didn't amplify. Shockley asked Bardeen, a recent graduate in physics from Columbia, to check his calculations. Bardeen did, and suggested that perhaps electrons were being trapped at the surface of the crystal, which would prevent them from seeing the calculated effect of amplification. Next, Brattain preformed experiments that confirmed Bardeen's explanation. So they were, after all, on the right track. They needed to play with the structure of the artifact, trying other configurations. One worked, providing a small amplification.

So a new kind of "electronic valve" was almost invented, but the amplification was too small. The group continued to invent alternative arrangements—one of which was to place two closely spaced gold strips on the surface of a doped germanium crystal and an electrical connection on its base. Relative to that base, one gold strip (called the emitter) had a positive voltage, and the other (called the collector) a negative voltage. They applied a signal between emitter and base, and produced amplification in the voltage between the collector and the base. The holes introduced into the germanium from the emitter went not to the base, but to the collector—this added to the positive current flowing from the base to the collector. It was, indeed, amplification. It meant this new device class, which they called a "transistor" could do everything the old electron tube could do—and with much less power dissipated as heat.

AT&T acquired a basic patent for the transistor and made transistors for its own use. AT&T licensed others to make them for other uses. At that time, AT&T was a legal telephone monopoly in the United States and was prevented from entering other businesses. In 1956, Shockley, Brattain, and Bardeen were awarded the Nobel Prize in physics. The electron vacuum tube began the first age of electronics, and the transistor began the second age of electronics. This is change from tubes to transistors is an illustration of a technology discontinuity.

We see, in this illustration, the intention to anticipate new technology by seeking a substitute for the electron vacuum tube. The basis of this anticipation was the vision of the need and the exploration of a scientific phenomenon (semiconducting materials) that might provide the opportunity for new basic invention.

TECHNOLOGY S-CURVE

The germanium transistor was an example of a radical innovation, and it set off a long period of further innovation. The pattern of this further innovation showed a general feature of technological progress, which has been called the "technology S-curve."

The rate of technical progress in any technology can be summarized by a technical performance whose increase implies greater usefulness. For example, in the new germanium point-contact transistor, an important performance parameter for amplification was the frequency range over which the transistor can provide useful amplification. When AT&T provided licenses for the transistor invention, a new Japanese company which later was to be named Sony purchased a license. But Sony engineers found that the AT&T version of the germanium transistor did not provide sufficient frequency range of amplification over the audio frequencies to be useful in a pocket radio. Sony researchers reinvented the transistor (from a "pnp" form to an "npn" form) to increase the frequency range of amplification. With this increase in frequency range, Sony was able to produce pocket radios, with which it began rapid growth toward becoming a global electronics firm.

Progress in a new technology can be measured by technical performance parameters whose increase implies greater utility of the technology.

When one plots the size of a technology performance parameter over time, one usually finds a similar form for most technologies. This pattern is (1) initial exponential growth in progress, (2) linear growth in progress, and (3) finally, asymptotically leveling off to little or no progress. Figure 11.1 depicts this pattern, which looks like a "lazy S"—hence its name, "technology S-curve."

Historically, most incremental technological innovation progress after a basic invention has followed this form. At first, all new basic inventions for a new technology show poor performance, are awkward and dangerous to use, and are costly to produce. Yet the opportunities for technical improvement begin as inventors and engineers seek ways of overcoming the limitations of the original invention. There is usually a rapid flush of new ideas that provides exponential increase in performance. All these new inventions in the exponential beginnings of a new technology are usually of the "trial and error" kind: "This is a new idea, let's try it to see if it works." For example, the Wright brothers' first airplane flew only 100 feet for 10 seconds

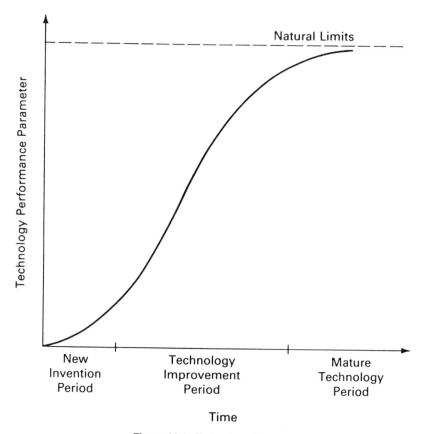

Figure 11.1. Technology S-curve.

at a 10-foot elevation in 1904. Yet only ten years later, in 1914, military planes provided observation duties at speeds of 75 miles an hour for ranges of 300 miles at altitudes of 5000 feet.

Eventually, and rather soon, all the obvious ideas get tried. Further progress in the new technology gets harder. Thus begins the linear phase of technology progress on the S-curve. Research scientists must now get down to better understanding and modeling of the phenomena on which the technology is based. Science understands and explores the materials and processes and creates models for predicting their behavior. With this refined understanding, engineers can invent new improvements to the technology. Since taking the time to research and understand and then to invent is slower then mere "trial and error" methods, research progress, in time, slows to linear from exponential. It is slower, but without such research, there is no further progress in the technology.

Can technical progress occur forever in a technology on a given phenomenal base? The answer is no. Nature is finite in any one phenomenon. As technological progress approaches the finiteness of the natural phenomenon on which it is based, then technical progress is limited by the finiteness of this natural phenomenon. This is called the "natural limit" to a technology, based on a given natural phenomenon.

The rate of progress in any new technology ordinarily follows an S-shaped curve, with an initial exponential rate, slowing to a linear rate, and turning off toward a natural limit.

Case Study: Inventing the Silicon Transistor

This case study illustrates another way in which technology progresses— through change in the science base (natural phenomenon) of the technology. The historical setting of this case study is about ten years after the original invention of the germanium transistor. The germanium transistor suffered from another serious technical performance problem: sensitivity to temperature. What good would transistorized electronics be if their circuit performance changed dramatically from the heat of day to the cool of night, from the hot of summer to the cold of winter? Many electronic engineers appreciated the new technology but yearned for a more reliable, less temperature-sensitive version of the transistor. The obvious route for most researchers was to try to make a transistor not from germanium, but from its sister element, silicon.

One of the groups looking for a silicon version of the transistor was based in a small U.S. company, Texas Instruments (TI). In 1952, Patrick Haggerty was president of TI, which was a maker of seismographic detection instruments, sold to oil companies for use in oil exploration. TI's instruments then used electron vacuum tubes, but needed to be portable and rugged and use little power. It was obvious to Haggerty that transistors would be a desirable replacement for tubes in TI's products.

Haggerty assigned to Mark Shepard, one of his employees, a research project to develop a germanium transistor that could be sold for $2.50 at that time. Shepard developed it, and TI produced a pocket radio with it in 1954. But TI did not follow through with this product; and Sony very soon, introduced its own independently developed germanium transistor and pocket radio and proceeded to commercially exploit this new product.

Still, Haggerty knew that the germanium transistor needed to be replaced. He hired a physicist from Bell Labs, Gordon Teal. Teal had been researching silicon, which was brittle and difficult to purify as a material for making transistors. Haggerty told Teal and another researcher, Willis Adcock (a physical chemist), to develop a silicon transistor. Many other research groups were seeking the silicon transistor; it was not an easy device to make. But Teal and Adcock did it.

In May 1954, Teal took the new silicon transistor to a professional conference, where he listened to several speakers tell of their difficulties in trying to make a silicon transistor. What he heard made Teal happy, for, as yet, no one else had succeeded. When Teal's time to speak came, he stood before the group and announced: "Our company now has two types of silicon transistors in production. . . . I just happen to have some here in my coat pocket" (Reid, 1985, p. 37). Teal's assistant came onto the speaker's stage carrying a record player, which used a germanium transistor in the electrical circuit for the amplifier. The tiny germanium transistor had been wired visibly outside the record player with long leads. Teal plugged in the player, put on a record, and started playing music. Next, the assistant brought onto the stage a pot of hot oil, and set it on the table beside the record player. Teal picked up the connected germanium transistor and dramatically dunked it into the hot oil. Immediately, the music stopped as the germanium transistor failed in the oil's hot temperature. Then, Teal picked up one of the silicon transistors, which he had earlier taken from his pocket and placed on the demonstration table. Teal took a soldering iron and replaced the germanium transistor with the new silicon transistor. The music from the record player sounded again. Then Teal picked up the silicon transistor and dunked it into the pot of hot oil, as previously he had with the germanium transistor. This time, the music did not stop. The silicon transistor could stand the heat. Texas Instruments had done it. Finally, a useful transistor—the silicon transistor—had been invented.

CHANGING THE NATURAL PHENOMENA UNDERLYING A TECHNOLOGY

We recall that technologies are composed of inventive logics to manipulate states of natural phenomena. Therefore, the phenomenal base of a technology is the nature manipulated by the logic of the technology. Technologies can be altered by changes in either (1) the logic schema or (2) phenomenal bases. In the above illustration, the silicon transistor used a similar logic schema to that of the germanium transistor (base, emitter, and collector) but a different natural phenomenon—silicon matter substituted for germanium matter.

Every technology S-curve depends on a phenomenal base. When a new phenomenon is substituted, a new technology S-curve begins.

When technology progresses by substitution of phenomena, then the overall technology S-curve is a composite of the underlying technology S-curves. Figure 11.2 is an illustration of a composite set of technology S-curves, due to the progress in the technology by substituting different natural phenomena.

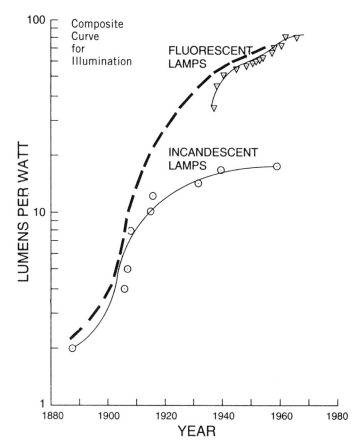

Figure 11.2. Composite technology S-curves based on different phenomena.

Case Study: Inventing the Integrated Circuit Chip

The transistor's progress from germanium to silicon altered the morphology of transistor circuits, but not the logic. This case study illustrates how progress in a technology can occur from altering the logic scheme of the technology without altering the phenomenal base.

The historical context is the middle of the twentieth century, after the invention of the transistor enabled more complex circuits to be designed because of the smallness and less power consumption of the transistor, compared to the older electronic tube. In the late 1950s, electrical engineers were using the new silicon transistor for many advanced electronic circuits. The transistors were so useful that a new problem arose in electronics: The new complex circuits engineers then could dream up might require so many transistors that no one could physically wire them together with the then-current technology of soldering transistors into printed circuit boards. A new natural limit was on the horizon for transistorized electronics—how many could be physically wired together. In 1958, this limit was on the mind of two researchers: Jack St. Clair Kilby at Texas Instruments and Robert Noyce at Fairchild.

Kilby had grown up in Kansas. As a child, he loved technical things and built a ham radio set using electron tubes (Reid, 1985). In September 1941, he went to the University of Illinois to become an electrical engineer. But three months later, the United States was at war with Japan, and Kilby joined the U.S. Army. He ended the war as a sergeant, assigned to an Army radio repair station on a tea plantation in India. He returned to college, graduated as an electrical engineer, and went to work at a firm named Centralab, located in Milwaukee.

One evening, he attended a lecture at nearby Marquette University given by John Bardeen, one of the inventors of the transistor. The new technology astonished Kilby, and he began reading everything he could find on the solid-state technology. In 1951, Bell Labs announced the licenses, and Centralab purchased one for $25,000. Contralab sent Kilby to Bell Labs for a five-day course in semiconductors. From 1951 to 1958, Kilby designed transistorized circuits for Centralab. As these circuits increased in complexity and numbers of transistors, Kilby saw the need for a technical breakthrough beyond hand-wiring transistors together.

Centralab did not have the research capability that Kilby would need to tackle the problem. In 1958, he sent his résumé to several larger firms; one went to Willis Adcock of Texas Instruments. Adcock hired Kilby, since TI was also worried about the integration problem of transistors. Adcock had a "micro-module" research project going at TI to approach that problem, to which he assigned Kilby. But Kilby didn't like the TI approach. One thing was clear to Kilby: Since TI led in silicon technology, a good solution to transistor integration should use silicon technology—if it could. Kilby's idea was: If transistors were made of silicon, could not the other components of a circuit, such as resistors and capacitors, also be fabricated on silicon? Why not make the whole circuit on a silicon chip?

It was an inventive idea, for then the technology for fabricating resistors used carbon as the phenomenal material and plastics and metal foil for capacitors. On July 24, 1958, Kilby wrote down in his lab notebook what he called a "monolithic idea": "The following circuit elements could be made on a single slice of [silicon]: resistors, capacitors, distributed capacitors, transistor" (Reid, 1985, p. 38). He entered rough sketches of how to make each component by properly arranging silicon material.

Kilby then showed his notebook to Adcock, who said that if he could demonstrate that the idea worked, an R&D project to develop it would be authorized. Kilby carved out a resistor on one silicon chip, carved a capacitor on another, and wired the two chips together. They worked, showing resistance and capacitance in the circuit. Adcock gave his approval to build a whole circuit on a chip. Kilby chose to build an oscillator circuit—one that could generate a sinusoidal signal.

On September 12, 1958, Kilby was ready. A group of TI executives assembled in the lab for a demonstration. The circuit on the chip was hooked to an oscilloscope, which would display the form of the signal from the circuit. Kilby switched on the new integrated circuit on the silicon chip, and it worked—a wiggling line appeared on the screen, the sinewave form of an oscillator circuit: "Then everybody broke into broad smiles. A new era in electronics had been born" (Reid, 1985, p. 39).

Sometimes, a basic invention is made independently about the same time by another inventive mind aware of the technological needs and opportunities. This happened with the invention of the semiconductor integrated circuit (IC) chip. That same summer, in 1958, Robert Noyce, who was a physicist and then-president of Fairchild Semiconductor, was also worrying about the problem of transistor integration. Fairchild was a new company manufacturing transistors (begun by one of the inventors of the transistor, Shockley, after he left AT&T). The following winter, on January 23, 1959, Noyce conceived a solution and wrote down in his lab notebook his monolithic idea: "It would be desirable to make multiple devices on a single piece of silicon, in order to be able to make interconnections between devices as part of the manufacturing process" (Reid, 1985, p. 41).

Both Texas Instruments and Fairchild Semiconductors filed for a patent on the IC chip. Years of legal argument followed about who had the right to the patent. In 1966, several semiconductor chip manufacturing firms met together; TI and Fairchild agreed to grant licenses to each other and share in inventing the device. Later, Kilby and Noyce both received (in addition to many other prizes) the National Medal of Science for inventing the semiconductor IC chip.

CHANGING THE INVENTIVE LOGIC OF A TECHNOLOGY

One can see in the preceding illustration that the key inventive idea of the IC chip was in the form of the logic of arrangement of the technology, and not in the form of a change of phenomenal base. The transistors on the IC chip were still made of silicon. The inventive idea was to also make and connect other electrical components also on silicon. This change in logic was powerful, as was the previous change in phenomena in the invention of the silicon transistor, where the phenomenal base was changed.

Therefore, in anticipating technological change for a technology S-curve, one must ask whether change can occur from different phenomenal bases or from different schematic logics, or both.

EXTRAPOLATING A TECHNOLOGY S-CURVE

Since a technology S-curve is a pattern for extrapolating technical progress, the fitting of the S-curve to a particular technology requires determining the two "inflection" points of the graph. (An inflection point on a curve is a point at which the second derivative to the curve changes sign.) In the initial exponential region, the S-curve is turning counterclockwise; at the first inflection point, the S-curve turns and begins pointing in a straight line. After a linear rate of change for a time, a second inflection point occurs when the rate of technological progress turns asymptotically toward zero.

If one can determine or guess these two inflection points (when and at what level of progress), then one can mathematically fit an S-curve form to data about technical progress. This is how a technology curve can be used for forecasting.

The first inflection point occurs when trial-and-error invention ends and (1) when research must begin for incremental progress or (2) when research for a next-generation technology is implemented. Incremental technical progress does not basically alter the phenomenal base or schematic logic of a technology, but refines one or both of them. Next-generation technology progress basically alters the phenomenal base or schematic logic, or both. The second inflection point occurs when the natural limits for manipulating the phenomenal bases of a technology are approached. Then, incremental research progress will produce smaller and smaller benefits because nature is limiting the basis for progress.

But finding or guessing where the inflection points are very difficult. In practice, it is usually best to try to figure out what the natural limit will be.

Case Study: Progress in Semiconducting Chips

This case illustrates the first instance in the history of technical progress whereby the rate of progress occurred as a sequence of exponential phases—because technical progress occurred in IC chips through next-generation technology research and implementations. The historical context is the period from 1960 to 1990, in the early years of technical progress in production of IC chips, when was new manufacturing processes were devised to achieve increasing densities of transistors on a chip.

The technology performance parameter is the number of transistors that could be fabricated on a chip. Since transistors worked as electronic valves, and at least one valve is usually needed in a circuit for one logic step, increases in the numbers of transistorized valves on a chip correlate with increases in the complexity of functionality for which a chip can be used. This is what a technology performance parameter should do—correlate an increase in parameter with an increase in functional performance.

In 1970, the feature size (the size of the components) of a transistor was about 12,000 nanometers wide. In 1980, feature size was down to 3,500 nanometers, and in 1990, 800 nanometers. In 1997, feature size was down to 300 nanometers. (One nanometer is one billionth of a meter; for comparison, a human hair has the width of about 100,000 nanometers.)

This reduction in feature size was allowing for the production of chips with greater densities of transistors. For example, by 1995, the transistor density in Intel CPU chips had followed this scale:

Year	Intel Chip	Number of Transistors	Word Length
1971	4004	2,300	4 bit
1974	8080	6,000	8 bit
1978	8086	28,000	16 bit
1982	80286	134,000	16 bit
1985	80386	275,000	16 bit
1989	80486	1,200,000	16 bit
1993	Pentium	3,100,000	32 bit
1995	Pentium Pro	5,500,000	32 bit

(Source: Intel Corporation)

These changes in size reduction and transistor density had been anticipated by the semiconductor chip industry. For example, in 1983, researchers were forecasting technical progress as summarized in Figure 11.3.

Note that, in Figure 11.3, the vertical axis of technical performance is on a log scale, so that the whole chart is in semilog format. A straight line of a semilog chart indicates exponential increase on the vertical scale. Therefore, region A is one exponential region of increase in chip density; region B is a second exponential region; and region C is also an exponential region. Since the five data points, 1960, 1972, 1979, 1980, and 1981, are actual historical data, they show that technical progress in IC chips occurred through next-generation technology innovations rather than incremental innovations.

After the innovation of the IC chip in 1960, the densities of chips increased to hundreds of transistors on a chip by 1970; this was called "middle scale integration" (MSI). After 1972, densities of transistors on a chip leapt to thousands of transistors, or large scale integration" (LSI). By 1981, densities of transistors increased to tens of thousands of transistors on a chip, or "very large scale integration" (VLSI). During the 1980s, chip densities increased to hundreds of thousands. As the 1990s began, chip densities were in the millions of transistors on a chip: ultra large scale integration (ULSI). Where would be the natural limits of silicon chip technology?

In the 1983 meeting on International Electron Devices, James Meindl,

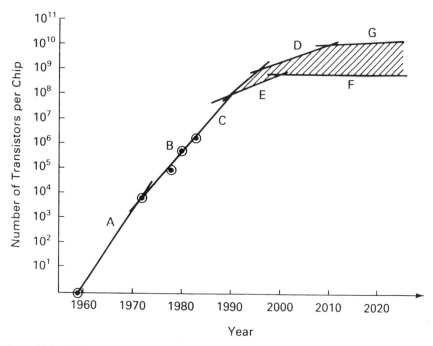

Figure 11.3. Technology progress in IC chip density. (From A. L. Robinson, "One Billion Transistors on a Chip?", *Science,* Vol. 223, January 20, 1984, p. 268. Figure courtesy of J. D. Meindl, Stanford University, Stanford, CA 94305. Copyright 1984 by the AAAS.)

then of Stanford University, predicted: "It could happen by the turn of the century" (Robinson, 1984, p. 267). At that meeting, there was an anticipation that the density limits could be in the billions of transistors per chip.

The limit would be in the tiny size of the fabricated features on a chip. IC chips, through the 1980s, were produced with light exposure to define features on a chip. Smaller size could be reached with X-ray exposure. But the shorter the wavelength of X-ray exposure, the more accidental damage to the features will occur. The ultimate limit to transistor density on silicon chips will occur from the technical tradeoff between fineness of feature definition and material damage from the exposing radiation.

What would be the "natural limit" to transistor size? Since transistors use electron current flow for operation, the natural limit would be making transistors operate with a single electron. In the mid-1990s, research laboratories were experimenting with trying to create a single-electron transistor, using only one electron to move in and out of a storage area for a binary-state transistor. This would provide an ultimate natural limit on transistor size:

> [S]cientists around the globe are making a furious assault on the last frontier of electronics. . . . They are striving to create transistors that work by virtue of the movement of a single electron.
>
> —(Broad, 1997, p. C1)

One prototype of a "single electron" transistor was to vary the "metal oxide silicon field effect transistor" (MOSFET) technology. In this technology, electrons flow from source to drain electrodes through a "channel" controlled by a gate electrode. The idea for a single-electron MOSFET is to make a small "island" in the channel that will only allow only one electron at at time to pass through the channel. Such "small islands" are also called "quantum dots" and operate according the the laws of quantum mechanics (which specify that only one electron can occupy one quantum state—i.e., Fermi statistics). For example, in 1996, research teams at NTT's Basic Research Laboratory in Atsugi, Japan, and at the University of Minnesota in Minneapolis each demonstrated such transistors with islands only 5 to 10 nanometers in size—small enough to alllow only a few electrons at at time through the channel (Service, 1997).

ANTICIPATING THE NATURAL LIMITS OF A TECHNOLOGY

The ability to estimate the natural limits of a technology lies in the detail of understanding of the natural phenomena of a technology—the science bases of the technology. Science provides detail through the quantitative modeling of the physical processes of the phenomena that a technology uses. These models allow predictions of behavior of the phenomena. The more accurate the scientific models of technologically useful phenomena, the better we are able to forecast natural limits for the technology.

In practice, to obtain such forecasts, a manager of technology needs to consult the research community studying the science bases of the technology.

FROM FORECASTING TO PLANNING TO INNOVATION

Forecasting technological change is an effort to foresee invention even before it occurs. But, on the way to invention, forecasting is not sufficient. One also needs to plan a new technology and perform the research necessary to invent it. The germanium transistor, silicon transistor, and IC chip were all examples of planned research and invention.

Planning technological change provides the basis for supporting efforts of invention. The combination of both anticipating and planning technological change is usually called "technology strategy."

TABLE 11.1   From Forecasting to Innovation

MOT Terms	Include the Following Concepts
Forecasting	Anticipating
Invention	Creating
Strategy	Anticipating, planning
R&D	Anticipating, planning, creating
Innovation	Anticipating, planning, creating, implementing

The anticipation, planning, and support for invention is what we earlier referred to as an "R&D system." When one does all these together—anticipation, planning, invention, and implementation—this is called "technological innovation."

Table 11.1 summarizes how these different terms are used in technology management.

RATES VERSUS DIRECTIONS OF TECHNOLOGICAL CHANGE

The "technology S-curve" summarizes the *rate* of change for a technology, but does not indicate the *direction* of change—how a technology system evolves. Although the patterns of the rate of change of technologies are similar for different technologies, the *direction* of change in a technology will be different for different technologies. To plan the directions of change for a specific technology, one needs another concept—"morphological analysis." What is similar about the direction of change across all technologies is the method of analysis and planning for change in a technology as a system.

We recall that all technologies are open functional systems, and that a technology system is a generic configuration of the open system. We further recall that technologies are a mapping of a *schematic logic* that expresses the functional transformation against as a *sequence of phenomenal states,* which provide the natural basis of the technology. We called these two aspects, logic and phenomena, the "schema" and "morphology" of a technology system. The inventive creation of a new technology requires devising both a schema and a morphology, and a one-to-one mapping between these. This understanding of technology (as a configured functional system with schema and morphology) can be used for technology planning.

Case Study: Progress in Vacuum Technologies

This case study illustrates the long-term historical progress of a technology through changes in its configuration. The historical period covers two hundred years of progress in vacuum technology.

The invention of vacuum technology began in the 1700s, when scientific experiments were performed by placing two copper hemispheres together and pumping out the air within the sphere by means of a leather bellows. This produced a partial vacuum within the sphere, and the force due to atmospheric pressure could be demonstrated by trying to pull the two hemispheres apart. It was very hard to do this, since the atmosphere was placing about 14 pounds per square inch of pressure all around the sphere.

The first useful vacuum pumps were mechanical pumps pushing a piston in a tightly sealed cylinder to force air out of an enclosed area. These were innovated in the 1800s. (The battery-operated tire pumps one buys in auto-parts stores to pump up an automobile tire works in this configuration.) By the 1940s, the best vacuum a mechanical pump could attain was a pressure of about 10^{-8} millibars.

In the 1950s, a new configuration of vacuum pump was invented to take the vacuum to lower pressures; this was the "diffusion" pump. The diffusion pump used a different physical phenomenon

In the 1960s, a new configuration was invented, the "ion pump," to reach lower pressures. In the late 1960s, a new configuration was invented, the "cryogenic pump," to attain even lower pressures. Each of these different kinds of pumps operates on different physical principles (phenomena and morphologies). Consequently, to reach lower and lower pressures in vacuum pumps, new configurations need to be invented:

Mechanical pump: 10^{-8} millibar pressure

Diffusion pump: 10^{-9} millibar pressure

Ion pump: 10^{-11} millibar pressure

Cryogenic pump: 10^{-13} millibar pressure

In the development of these succeeding devices to advance vacuum technology, several other devices had to be invented, in addition, to make them all possible. For example, essential to the invention of the ion pump was the invention of the ion gauge and the development of elastomer sealants and impermeable helium glass. The cryogenic pump required the development of additional low-temperature technologies and the mass spectrometer gauge.

One sees in this illustration that major progress in a technology often requires inventing new system configurations along with other complementary technologies for the new system configuration. A new system configuration of a technology system is an example of what we have earlier called a next-generation technology.

MORPHOLOGICAL ANALYSIS FOR ALTERNATIVE CONFIGURATIONS OF A TECHNOLOGY SYSTEM

The logic for exploring alternative morphological configurations of a technology system was devised by an engineer, F. Zwicky, who called this logic "morphological analysis." What he meant by this name was a systematic analysis of alternate structural (morphological) configurations. Zwicky wrote:

> An exact statement is made of the problem which is to be solved. For instance, we may wish to study the morphological character of all modes of motion, or all possible propulsive power plants, telescopes, pumps, communication, detection devices, and so on. If one specific device, method or system is asked for, the new [morphological] method immediately generalizes the inquiry to all possible devices or systems, which can provide the answer to a more general request.
>
> —(Jantsch, 1967, p. 175)

For example, recall the earlier case study of the invention of the helicopter. Flight, as a technology, had been envisioned in two different configurations: airplanes, with fixed wings, and helicopters, with rotating wings. In Zwicky's terminology, these two different system configurations of the technology of flight, were called two different "morphologies" (or structures) of the technology. This is because the logical schema of the technology of flight (power-on, lift, fly, control, land) can be mapped onto two different structural configurations: fixed-wing structures or rotating-wing structures.

Zwicky's procedure was relatively straightforward. He suggested starting with one configuration of a technology system, abstracting its structural features, and then generalizing on all logical alternatives in such features. As an example of this procedure, Zwicky used the general structural features of a jet engine technology:

1. Intrinsic or extrinsic chemically active mass
2. Internal or external thrust generation
3. Intrinsic, extrinsic, or zero-thrust augmentation
4. Internal or external thrust augmentation
5. Positive or negative jets
6. Nature of the conversion of the chemical energy into mechanical energy
7. Vacuum, air, water, earth
8. Translatory, rotatory, oscillatory, or no motion
9. Gaseous, liquid, solid state of propellant
10. Continuous or intermittent operation
11. Self-igniting or non-self-igniting propellants (Jantsch, 1967, p. 176)

Having identified the main structural features of jet engine technology, Zwicky next put all the alternative features together in possible combinations that included at least one feature from this list of 11 features. Zwicky computed that there would be 36,864 possible combinations, but many combinations would have no technical interest. Zwicky performed this analysis in 1943, using fewer parameters, and reduced the interesting possibilities to 576 alternative configurations. Later, Erich Jantsch noted that Zwicky had correctly anticipated five possible combinations that were developed. (In fact, in 1943, two of Zwicky's combinations were the secret German devices that were being used to bomb England: the pulse-jet powered aerial bomb (V-1) and the liquid-fueled rocket (V-2).) Jantsch noted a historically interesting footnote:

> One may recall, in this context, that the fatal failure of Lindemann, Churchill's scientific advisor, to recognize the potential of the V-2 even when he was shown photographs [Lindemann had asserted: "It will not fly"] is plausibly explained by his exclusive preoccupation with solid propellants, stubbornly rejecting the idea of liquid propellants.
>
> —(Jantsch, 1967, p. 178)

What Zwicky's approach points out is that morphological alternatives to a configuration of a technology system can be envisioned by:

1. Abstracting the salient features of the technology structure
2. Generalizing on logical alternatives in each feature
3. Taking combinations of each alternative feature to see different configurations of the technology system
4. Focusing attention on technically interesting configurations

While Zwicky's specific approach is very general, it is also clumsy and combinatorially large. So Zwicky's approach is seldom used directly, but all technology planning is based upon some form of morphological analysis. The key idea is that:

Places in a technology system for research to improve the system can be identified by looking at all its structural features.

Case Study: Xerox Corporate Research Organization in 1981

This case study illustrates the organization of a corporate research laboratory in a major corporation, using morphological ideas. The historical context is in the early 1980s in the United States, when corporations were still investing strongly in basic research.

In 1981, Xerox's business was office copiers, based on the xerography invention on which Xerox was founded and developed. Xerox dominated the office copier market, but was facing both maturity of the market and, at the same time, competition from competitors in the low-end copier market. Earlier, in the 1970s, Xerox management had decided to explore new the new computer technologies and their connection into the copying market. (Xerox had bought a mainframe computer company in the 1960s and then sold it in the early 1970s when it failed to effectively compete against IBM's dominance of the mainframe market.) Xerox decided to reenter the computer business based on new technology in computers, and set up the Palo Alto Research Center (PARC) laboratory to pioneer new computational ideas for Xerox's technical vision of the "paperless office."

Xerox then supported three research laboratories, two in technologies for the xerography business and one for the hoped-for computer business: the Webster Research Center, Xerox Research Centre Canada, and Palo Alto Research Center.

The Webster Research Center was the oldest laboratory, established around the technology system of xerography. It specialized in marking and imaging technologies, as these are the core technologies underlying xerographic copying systems. Xerography was the technology on which Xerox was built, with copiers then its main business. Then M. D. Tabak directed the center. The Xerox Research Centre Canada was focused in materials processing research, advancing the materials technologies also involved in xerography. R. H. Marchessault was then director of the Canadian center. Together, the Webster and Canadian centers covered the technologies for xerographic copying systems. The Palo Alto Research Center (PARC) had been set up to investigate computational systems, focusing on digital systems research. R. J. Sinrad was then director of PARC.

Comparing the research strategies of Xerox at two points in time, in 1972 and later in 1981, one can see that "morphological analysis" of the technologies relevant to Xerox's business underlay the strategies.

For example, in 1972, Xerox's research areas were mostly focused on xerography:

1. Photoreceptors—alloy and organic
2. Understanding the xerographic development process
3. Laser xerography
4. Solid-state lasers

There were two additional areas with the idea of computers as a future technology:

5. Optical digital storage
6. Distributed systems

And there was a general area for science:

7. Science base

In 1981, the research efforts of the laboratories still included xerography, but had broadened more over the computational vision. For xerography, the different morphological topics were grouped together:

1. Marking technologies and transducers
2. Materials

For computation, topics were expanded:

3. Optical storage
4. Programming distributed files
5. Human-digital interface
6. Microelectronics

And there remained a general area for science:

7. Science base

One can see that Xerox research over that decade kept the morphological emphasis on xerography technology, and the Webster and Canadian centers focused on this. At PARC, the research focused around computational technologies.

PARC's research was so successful that its computational system in 1981, then called "Altos," presaged the personal computer systems of the 1990s. In fact, Steve Jobs of Apple visited PARC in the early 1980s and used PARC's vision of computation to design the Macintosh series of personal computers that was to save Apple in the 1980s. It gave Apple a technology leadership position in personal computers for a decade.

This case is typical of the organization of the research areas of industrial research laboratories. The research areas are focused around the morphologies and logics of the technologies core to their current and future businesses.

TYPES OF TECHNOLOGICAL CHANGE TO BE PLANNED

We recall that different ranges of technology are used in a firm: products, production, and information. New technology for products and external services is focused either on improving existing products and services, creating new products and services, or creating new businesses. New technology for production is focused either on improving existing production systems or creating new production systems for new products, services, or businesses. New technology for improving information capability is focused on improving communication, management information systems, or delivery systems.

The three kinds of technologies—product, production, and information—may be interconnected. Production processes depend on the kinds of products produced; and operations depend on the kinds of production to be controlled and markets to be serviced. Accordingly, in logical order, technology planning for a new enterprise, should begin with product planning, and then move to production planning, and to operations and information planning. For an established enterprise, however, technology planning should first pay attention to the technologies that are changing most rapidly, either in product, production, or operations, and then ask what impact such new technologies can make on the enterprise.

PROCEDURES FOR PLANNING TECHNOLOGICAL PROGRESS

Although morphological analysis is important to envision alternative structural configurations of a technology, the specific procedure to plan technological progress has many additional steps. These include:

1. Technology audit
2. Competitive benchmarking
3. Customer needs and requirements
4. Technology barriers
5. Technology roadmap
6. Next-generation technology system

A technology audit is a procedure to systematically identify the core technologies of a business. Competitive benchmarking is a procedure to compare a company's technologies, products, services, and production capabilities with competitors' capabilities.

Identifying current and future customer needs and requirements looks to the application systems of the customer to envision the need for technological progress. Technology barriers are the points of technology morphologies and logics in current technologies that must be to improved to get from current technologies to envisioned future technologies. A technology roadmap lays out the paths from current technologies to future technologies if the technology barriers can be overcome.

A next-generation technology system is a research vision of what kind of research program could tackle the technology barriers in a technology roadmap.

TECHNOLOGY S-CURVE AND CORE-TECHNOLOGY MATURITY IN AN INDUSTRY

We recall that two forms of core technological maturity occur in any industry: (1) product standardization and (2) process standardization. In Figure 11.4, we can combine the charts for (a) industrial technology life cycle, (b) rates of product and process innovations, and (c) technology S-curves for key technologies of the product system.

Looking along the time dimension, we see that the major growth in a market in the applications growth phase begins when the design is standardized for the product. Eventually thereafter, the industry begins a "mature technology" phase for its product and the product becomes a "commodity-type" product, as opposed to a "high-tech" product. The technology S-curve for the core technology of the product levels off as product innovations stop.

During this time, however, technology innovation continues in the production process; a "commodity-type" product still uses high-technology, technological change, for its production. During this time, market volume can still grow, as production innovations reduce the cost of producing the product. As the process innovations decline, then the technology S-curve for the production processes levels off.

Only after both product and production innovations have declined does one have a totally "technology-mature" industry. Then, until a substituting technology occurs for the core technology of the product or for the core technologies of production, the technology-mature industry can continue at a market level determined by market replacement rates and/or market demographics.

SUMMARY

Being able to anticipate technological change is essential to exploiting a technical window of commercial opportunity. Progress in a new technology can be measured by technical performance parameters whose increase implies greater utility of the technology. Once a basic invention for a new technology has occurred, the pattern of further progress is: (1) initially, exponential growth in progress; (2) next, linear growth in progress; and (3) finally, asymptotically leveling off to little or no progress. This form is called a technology S-curve, and for a particular technology requires determining the two inflection points of the curve, linear and asymptote. The ability to estimate the asymptotic natural limits to a technology lies in the detail of

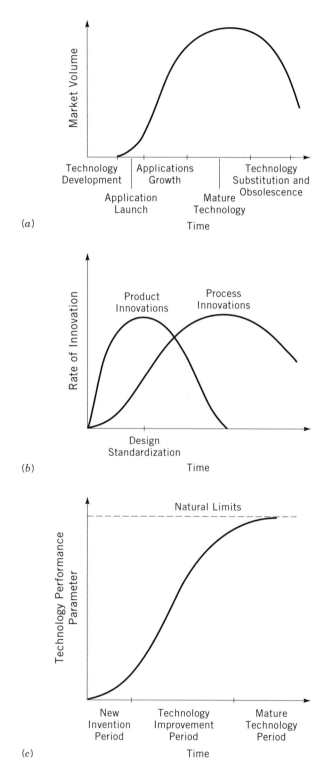

Figure 11.4. (*a*) Core technology industrial life cycle. (*b*) Rates of product and process innovations. (*c*) Technology S-curve for core technology of product system.

understanding of the natural phenomena of a technology—the science bases of the technology.

Technology planning goes beyond technology forecasting to establish the research necessary to make a technology forecast become a future reality. Technologies can be altered by changes either in morphologies or logic schema.

FOR REFLECTION

1. Think of a technology (device, system, or service) with which you are familiar. What are the measures of performance for the technology? Plot the change in performance over time. Did it take the shape of an S-curve? Did it jump from one curve to another? What are the natural limits to the technology, and on what are these based?

2. Obtain recent annual corporate reports from a company in a research-intensive industry, and then obtain material from this company on how its R&D activities are organized. Discuss how the research efforts reflect the company's vision of its current or future businesses.

____12
TECHNICAL PROJECT MANAGEMENT

CENTRAL CONCEPTS

1. Project management for strategic change or for operations
2. Types of technical projects
3. Logical stages in technical projects
4. R&D project returns
5. Estimating discounted returns
6. Project planning
7. Project scheduling
8. Project teams
9. Project management in bureaucracies

CASE STUDY

Pilkington's Float Glass R&D Project

INTRODUCTION

We have just reviewed techniques for anticipating and planning technological progress. The actual practice of creating technological progress occurs in discrete activities, such as R&D projects and design projects. Thus, technical projects are the forms of activities for implementing technological change.

We will now review the general practice of managing projects. We recall that there are three styles of management:

1. Bureaucratic management for running large organizations,
2. Entrepreneurial management for starting and building organizations
3. Project management for introducing finite change in an existing organization

Project management is the style of running technical projects because these projects are focused on *creating a specific incident of change.* Harold Kerzner emphasized that a project should have:

- A specific objective
- Definite start-stop dates
- Funding limits
- Resource requirements
- A specific customer (Kerzner, 1984)

We will now review the topic of project management, emphasizing its application to technical projects.

PROJECT MANAGEMENT IN MECHANISTIC OR ORGANIC ORGANIZATIONAL CULTURES

Project management in an organization may be focused on projects for either strategic or operational change. Lowell Steele emphasized that management concerns can focus on the shorter-term concerns of operations or the longer term concerns of strategy (Steele, 1989). Operations are the repetitive daily activities of a business that produce and sell products and create profit. Strategy consists of the periodic and nonrepetitive activities that create change in operations. Also, as we noted earlier, firms may be classified as to whether they produce large volumes of the same product for many customers or produce one customized product at a time for a few customers. Burns and Stalker called these 'mechanistic' or 'organic' forms of organizations.

In a mechanistic form of organization, multiple copies of the same product design are sold to many customers, and operations are focused on large-volume production. In this case, management of operations concentrates on the control and stability of repetitive activity. Change needs to occur only periodically in mechanistic culture firms, when operational procedures require changing; and such change is strategic. Project management that is focused on strategy for mechanistic kinds of organizations alters the policies and procedures of operations.

In organic organizational cultures, each product is customized to a specific customer and one-of-a-kind products are produced. Here, project management is the form of production, and project management is the operational form of management at the production level. Examples of organic kinds of business organizations are engineering consulting firms, architectural firms, construction firms, R&D firms, law firms, and medical practices. In organic firms, project management is the operational form of management and is not focused on strategic change.

TECHNICAL PROJECTS

Whether in mechanistic or organic organizations, technical projects are of three sorts: research, design, or operations. Research-based projects are usually called "R&D projects." Design-based projects are usually called "engineering projects." Operations-based projects are usually called "systems projects." R&D projects innovate new technologies, engineering projects reapply existing technologies; and systems projects develop new or improved operations. Each of these types goes through different kinds of stages.

R&D Projects

The stages of R&D projects are:

1. Basic research and invention
2. Applied research and functional prototype
3. Engineering design and testing
4. Product testing and modification
5. Production design and pilot production
6. Initial production and sales

Stages 1 through 3 are usually called "research," while stages 4 through 6 are called "development"; thus arose the name "research and development" (R&D). Each stage of innovating a new product is expensive, with the expense increasing by an order of magnitude at each stage. The management decisions to proceed from research to development are, therefore, very important.

Engineering Projects

The stages of engineering projects are:

1. Customer and needs identification
2. Product concept
3. Engineering specifications

4. Conceptual design
5. Detailed design
6. Prototype engineering design
7. Production design
8. Testing
9. Production

Engineering design uses working technologies, applying them to customer needs. Accordingly, engineering design begins with identifying customers and their needs. These functional needs must be formulated as engineering specifications, which the creative activity of conceptual and detailed design addresses. What emerges is a prototype engineered design of a product or process. This design needs to be modified for production and then tested.

Systems Projects

The stages of systems projects are:

1. Customer and function identification
2. System definition
3. System requirements and specification
4. System architecture design
5. System component design
6. System control design
7. System prototyping and programming
8. System testing
9. System implementation

Systems projects tend to emphasize operations, as opposed to the artifact orientation in traditional engineering projects. In fact, a device, such as an airplane, for which operations requirements are complex, may be called a systems project. Software projects that produce coding for operations are usually also called systems projects. Projects designing operational systems, such as transportation systems, communication systems, or information systems, are usually called systems projects as well.

In most system projects, known technologies are used and applied; therefore, systems projects usually begin with customer and functional identification. From this, a system must be identified with boundaries and specifications. System design activities then include design of architectures, components, and controls. Completed system designs must be prototyped and programmed, and then tested and implemented.

Case Study: Pilkington's Float Glass R&D Project

This case study illustrates the stages of an R&D project. The historical setting is in the 1950s, when an invention to radically change how to manufacture plate glass was substituted for existing technology throughout the glass industry. This is also an example of technical project management for strategic change of operations in an organization of "mechanistic" culture.

In 1952, Pilkington was the major glass manufacturer in the United Kingdom, controlling 90 percent of that market with annual sales of £113 million. At that time, the world's glass industry was oligopolistic—dominated by a few large producers, with major shares of their respective national markets. Some of the other major glass firms in the world were:

- In the United States, Libby-Owens-Ford (annual sales of £187 million)
- In Canada, PPG Industries (annual sales of £477 million)
- In Japan, Asahi (annual sales of £265 million)
- In France, Saint Gobian (annual sales of £436 million) (Layton, 1972)

Until 1952, flat glass was continuously manufactured in one of two ways. The first way was to draw glass upward as a ribbon from a bath of molten glass; this produced a low-cost, fire-polished glass, yet with considerable optical distortion, since the glass surface wavered as it was drawn upward. The second method was to roll cooling glass horizontally between rollers, which pressed the glass to the right thickness but left marks from the rollers; expensive polishing and grinding then had to be done, which produced optically better glass but at a much higher cost.

Such was the state of the technology when the young Alastair Pilkington, a cousin of the owning family, went to work for Pilkington. Alastair began by working in the sheet-glass division and became familiar with the inexpensive method of glass production, drawing glass upward. He then became production manager of the plate glass division and observed the very expensive process of grinding and polishing plate glass.

Invention. Alastair's inventive mind began thinking about alternative ways of producing plate glass. He saw the natural result of flatness when glass floated upward as it was pulled vertically—why not float it horizontally? Molten glass could be floated out on a bed of molten metal; he chose to try a hot bath of tin. Molten tin had a low enough melting point to remain molten as the glass solidified on it. Moreover, the tin could be kept free of oxide if the atmosphere were controlled. Alastair presented his idea to his superior, the production director of the flat-glass division, who took the idea to the company's board. The board recognized the potentially vast economic return if the idea worked, and approved a budget for an R&D project.

Applied Research and Functional Prototype. A small project group was created in 1952, consisting of Richard Barradell-Smith and two graduates, reporting to Alastair Pilkington. They built a small pilot plant, costing £5000 that produced 10-inch-wide plates. They found that, when glass cooled on the molten tin, it happened to be precisely 7 millimeters thick. (This happened due to a physical balance between the surface tensions and densities of the two immiscible liquids, molten glass and molten tin.) This was lucky because then, 60 percent of the flat-glass trade was in 6-millimeter-thick glass. By stretching the glass ribbon a little as it cooled, they could reduce the natural 7-millimeter thickness to the commercially desired 6 millimeters.

Engineering Prototype and Testing. Next the project group built a second pilot plant to produce 48-inch-wide plates (this was the commercial width). It cost £100,000 (five times more than the first 10–inch machine). These two machines and their research took five years, until 1957, until they were working properly. By then, the members of the research team thought they had perfected the production technique enough to try a production-scale plant.

Production Prototype and Pilot-Plant Production. In 1957, the R&D team was expanded to eight people, with three graduates from different disciplines. The group was moved into the plate glass factory and built a plant costing £1.4 million (14 times the cost of the engineering prototype).

The major purpose of research is to reduce technical risk before production-scale investment is committed. But technical problems always occur. At the plant scale of production, the first problem was that the control of the atmosphere had to be improved to prevent oxidation of the tin. Then, problems of flowing the large quantities of glass out rapidly onto the tin had to be solved. Also, there were problems at this scale of stretching the glass to the right thickness and pulling the cooled glass plate off the molten tin. At each scale-up, previously solved technical problems had to be solved again. (This is typical of engineering problems in production.) This problem solving phase of the development of the full-scale production process took 14 months, from 1957 through 1958, and produced 70,000 tons of unsalable glass.

Then, finally, all the problems were solved and salable glass was produced. Production was ready.

Initial Production and Sales. In 1959, the plant went into commercial production. Another plant was built in 1962, and another converted in 1963. The total expenditure on the development and commercialization of the new process was £1.9 million over ten years. Pilkington had innovated a commercial process to produce plate glass at the inexpensive cost of sheet glass. It obtained worldwide patents on the valuable process.

What business strategy did Pilkington choose for commercial exploitation over the world? All its competitors were vulnerable. The process was so radical in improved quality and lower cost, that the firm could take away any competitor's market.

Pilkington decided to license the new process to competitors in other countries, rather than trying to expand internationally. Several reasons went into the decision: the great amount of capital required for worldwide expansions, the location needs for plants—near supplies of iron-free sand and sources of soda ash, quarries for dolomite and limestone, and cheap fuel—all resources that competitors had lined up in their own companies. From 1959 to 1969, all producers purchased licenses from Pilkington. For a time, the license revenue provided Pilkington with a third of the company's profits; and after a period of time (generally, 17 years), however, Pilkington's patents expired.

R&D PROJECT RETURNS

All technical projects require investment to be commercialized. A technique for projecting the financial investment and return in a technical project is called the "project financial chart, as sketched in Figure 12.1. This chart consists of a horizontal time axis and a vertical financial axis. On the vertical financial axis are plotted (1) the cumulative profits from commercialization of a technical project and (2) the sales volume of products from the technical project. (If the technical project results not in a product but in production implementation or improvement, then the cumulative profits shown should be the contribution to profitability from the production project.)

A financial project chart graphs both the projected costs and returns of the project and sales volume over time.

When the finances of the project begin at a time t = zero on the horizontal time axis, then costs of the project increase down the negative vertical financial axis of the graph, until the point where the results of the project are first commercialized. If the results are a product, then this point is at the commercial launch of the project. At this point, sales of the product begin. As sales volume grows, then the project costs are covered by the profits from the sales until a "break-even" point occurs. At this time, the sales from commercialization of the project just offset the project investment costs.

Cumulative profits from the project continue to climb positively on the vertical financial axis as sales volume grows. If the sales of the project decline from product obsolescence, the cumulative profit will level off and end when the product is withdrawn from the market.

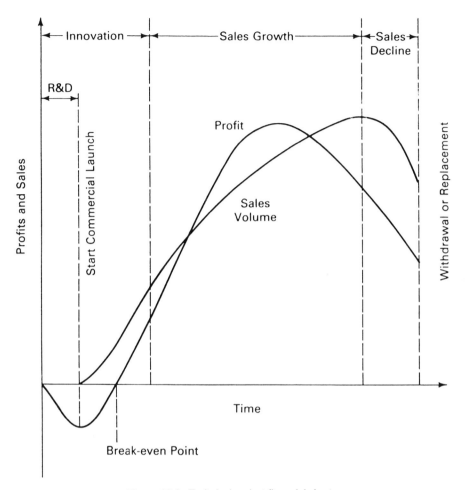

Figure 12.1. Technical project financial chart.

DISCOUNTED FUTURE VALUE OF PROJECTS

Another financial technique to value the future return from an R&D project is a discounted future calculation. If one has a choice of two projects, either of which can create a new product line, and there are limits to investment such that both cannot be developed simultaneously, then there are several criteria on which to choose between them, some about the business and some about the timing. A technique to compare the timing differences is to calculate what is called the "discounted future return rate" of the projects. This calculates not only the cumulative financial return from commercializing the project, but also the rate at which the return occurs. Returns occurring earlier can be seen as financially more valuable than later returns because earlier returns can be increased by reinvestment of the earnings. Thus, a financial return later for one project than an earlier return by another should be discounted by its later time of return when comparing their financial returns.

For example, a simple way to compare discounted future returns is to assume that an investment could be alternatively placed into a project or into an interest-bearing bank account. If the return on the project is x years in the future, this return can be compared to the alternative deposit in the account with its return compounded at a fixed interest rate over the same x years. What this comparison does is to see if the project investment can return more income than an alternative interest-bearing account.

This technique should be used cautiously, however, because some R&D investments are necessary simply to stay in business, and alternative forms of investment are not feasible. For example, in 1982, Robert Hayes and David Garvin cautioned about misusing discounting techniques:

> Highly sophisticated analytic techniques now dominate the capital budgeting process at most companies. . . . As these techniques have gained ever wider use in investment decision making, the growth of capital investment and R&D spending in this country has slowed. We believe this to be more than a simple coincidence. We submit that the discounting approach has contributed to a decreased willingness to invest. . . . Bluntly stated, the willingness of managers to view the future through the reversed telescope of discounted cash flow analysis is seriously shortchanging the futures of their companies.
> —(Hayes and Garvin, 1982, p. 71)

PROJECT MANAGEMENT PHASES

In the management of a technical project, several management phases occur. The project needs to be planned, implemented, and completed. This creates distinct management phases:

1. Planning phase
2. Staffing phase
3. Performance phase
4. Implementation phase

Project Planning Phase

The planning phase should include participation by all the potential stake-holders in the project: research and development personnel, marketing and production personnel, and finance personnel. Buy-in by relevant parties into the plan is important for the eventual success of the project. This buy-in formally results in a project plan and an allocation of resources to perform the project.

The project plan should succinctly state the goal of the project, the means and timing of the project, resources required, and the customer for the project:

- What is to be done
- Who will use it
- How and when it is to be done
- Who is to do it
- How progress will be monitored
- How the customer will get it
- What it will cost

Accordingly, a project plan should include the following sections:

1. Project objectives
2. Project customers
3. Project schedule
4. Project staffing requirements
5. Project review procedures
6. Project implementation plan
7. Project Budget

Project Staffing Phase

Once a project plan is approved by higher management and a budget allocated, the next project management phase is staffing and beginning the project. Project staffing requires:

1. Recruiting
2. Developing the subculture of the project team
3. Organizing the activities
4. Coordinating and liaison and developing shared supervision of team members between the project and the functional homes of the team members

The quality of the team will determine the quality of the project outcome. A technical project team needs to be technically multidisciplinary and multifunctional in business to cover both the technical and business dimensions of the project.

Project Performance Phase

Project monitoring requires:

1. Developing measures of technical progress in the project
2. Developing a modular task structure for organizing the work
3. Developing techniques to track technical progress (such as PERT charts, GANTT charts, etc.); see Figure 12.2

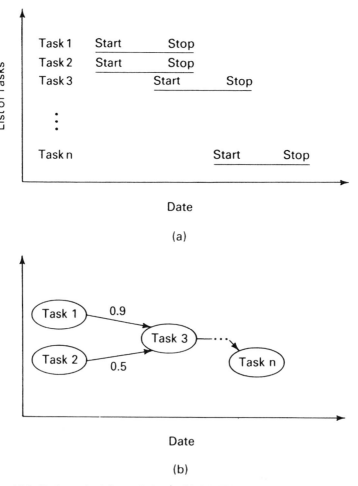

Figure 12.2. Project scheduling techniques: (*a*) GANTT charts and (*b*) PERT charts.

4. Developing modes for problem solving and debugging during the project
5. Developing a testing program for modular testing of the system
6. Developing proper documentation for the project as the project progresses

Sample project monitoring charts are shown in Figure 12.3.

Project Implementation Phase

Project completion requires:

1. Prototyping the project results
2. Demonstration of the prototype's performance
3. Transferring the completed project to the project's customer
4. Assisting the project's customer with the implementation of the next phase of the project product or putting it into use
5. Rewarding project team members for contributions to the project

Project success requires:

1. Achievement of technical performance
2. Results achieved within the window of opportunity
3. Achievement of results within a budget that can produce economic benefit
4. Results that satisfy the project customer
5. Effective rewards for successful performance to project personnel

PROCESS AND TECHNIQUES FOR PROJECT MANAGEMENT

The process within which projects are managed can be formalized when there are many projects, as in an organic organizational culture. For example, in an R&D laboratory, there are many projects, and a formal process for formulating, selecting, monitoring, and evaluating projects is usually established. Figure 12.4 illustrates the levels of management involved in R&D projects (modified from Costello, 1983). Senior management sets research priorities and business directions. Middle managers then define project requirements, around which they and technical staff generate research ideas. Technical staff then draft R&D proposals that are sent to middle and senior management, who select which projects to fund. Funded projects are performed by technical staff and monitored by middle management.

Senior and middle management review project portfolios periodically and

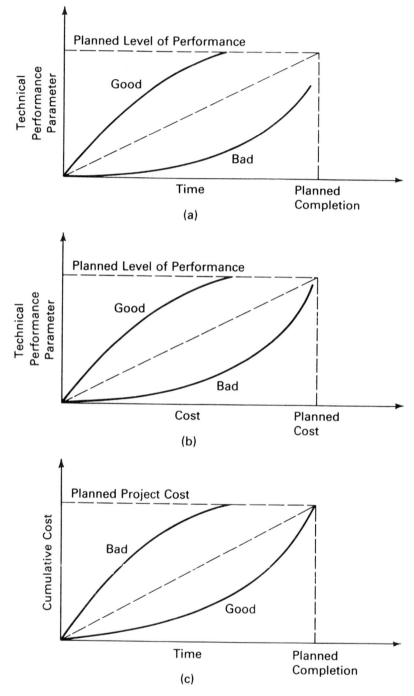

Figure 12.3. Project monitoring charts.

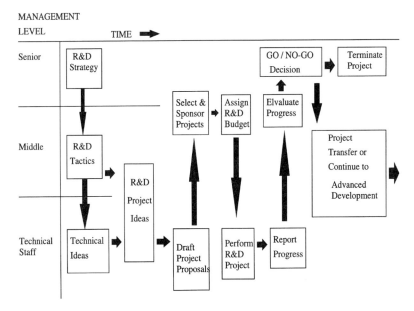

Figure 12.4. Formal R&D project process in a research lab.

select research projects to terminate or to continue into development and implementation, based on technical progress and commercial importance.

There are now many software aids for the time, budget, and resource aspects of project management, such as "project schedulers." Also embodied in some software are techniques for determining the priority of tasks and risks of completion, such as the Program Evaluation and Review Technique (PERT). Other forms of software, such as "groupware," facilitate the communication and recording of project documentation.

COMBINATION OF R&D, ENGINEERING, AND SYSTEMS PROJECTS

Sometimes, when new technology is being developed and embodied into new systems wherein artifacts are designed, a technical project may combine the R&D, engineering, and systems stages. This often happens in high-tech projects that are advancing technology by a dramatic step. In these projects, it is important to sort out technical risk from commercial risk and attend to each kind of risk in an appropriate and timely manner. Combined technical projects are inherently more risky and expensive than separate kinds of technical projects, because of the multiple kinds of uncertainties in both technology and commercial applications.

TECHNICAL RISKS

In innovating new technology, project management techniques should be adapted to technical risks. For example, Aaron Shenhar has emphasized identifying the degree of technical risk of the project:

> The technology used in projects is one of the main aspects which should receive special attention, since there are great differences among projects. Some projects use well-established technologies, while others employ new and sometimes even immature ones, and involve enormous uncertainties and risks.
>
> —(Shenhar, 1988, p. 2)

Shenhar suggested classifying a development project by the degree of risk in the technology being implemented:

1. Low-tech projects
2. Medium-tech projects
3. High-tech projects
4. Super high-tech projects

A low-tech project is one in which no new technology is used. A medium-tech project is a project in which some new technology is used within an existing system, but is a relatively minor change in the whole technology of the system. A high-tech project is one that uses only key technologies as components, but in which the integration of these technologies is a first-time achievement. A super high-tech project is one in which new key technologies must be developed and proved along with their integration into a new system.

Project planning for low-tech projects should use the experience of previous projects of the same type for estimating scheduling, budgeting, and staffing and for arranging for completion and transfer. Here, a PERT chart is very useful, wherein past experience can be used to estimate probabilities of risk within the chart.

For a medium-tech project, it is important to determine how much competitive advantage the new technology provides in the project: nonessential or essential. If nonessential, then a substituting technology should be within the scope of the design and ready for use if substantial problems develop with the new technology. If essential, then it is important to first plan and schedule the development and testing of the new technology before system integration is far advanced. The scheduling of system development should not begin until confidence is demonstrated that the new technology will perform properly and on time, for the length of the project schedule. Here, probabilities in a PERT chart should be used only for established technologies and not for the new technology.

For a high-tech project, it is important to plan and schedule the system integration in a modular, subsystem fashion, so that subsystems can be thoroughly tested and debugged before final system integration and assembly occurs.

For a super high-tech project, the planning of the project should be in the form of a series of "go/no-go" decisions. Each stage of the project development should consist of first developing and testing a key technology before the full project system integration program is launched. Whenever a key technology fails, a no-go decision should be made to keep the full-integration program on hold until either the key technology can be made to work or an adequate substituting technology can be implemented.

In summary, for low, medium, and high-tech projects, development can be scheduled for fixed dates (with appropriate regard to project design for testing, in a modular fashion, a new-technology component or subassembly); but for a super high-tech project, full system development should not be scheduled until all key technologies have been successfully proven or substituted.

PROJECT SELECTION

As we noted earlier, in the practice of industrial research, project portfolio selection is generally ordered around morphological areas of the core technologies of the firm. But within morphological areas, there is still often a problem of choosing an appropriate portfolio of R&D projects in a given area. Then projects are compared on their technical merit and commercial potential. There has been a long history of attempts to provide quantitative models for doing this. They have been called formal models for project selection, and periodically there are reviews of this topic.

For example, in 1994, Robert L. Schmidt and James R. Freeland reviewed progress in formal R&D project selection models (Schmidt and Freeland, 1994). They grouped models into two approaches: "decision-event" or "decision-process." The decision-event models focus on outcomes: given a set of projects, the model determines the subset that maximizes an objective. But the history of actual use of these models is poor, as Schmidt and Freeland state:

> Classical R&D project selection models have been virtually ignored by industry. . . . This fact became apparent in the early 1970s.
> —(Schmidt and Freeland, 1994, p. 190).

The decision-process approach focuses on facilitating the process of making project selection decisions, rather than attempting to determine the decision. Schmidt and Freeland suggest that formal project selection models should assist in coordinating the decisions about selecting and monitoring a project portfolio. Other reviewers of the formal project selection literature agree. For example, Carlos Cabral-Cardoso and Roy L. Payne wrote:

> Although a large body of literature exists about the use of formal selection techniques in the selection of R&D projects . . . much skepticism exists about their value.
>
> —(Cabral-Cardoso and Payne, 1996)

They also argued that a judicious use of formal techniques can improve communication and discussion about project selection. What constitutes judicious use is the problem. All formal decision methods of R&D project selection techniques require estimation of probabilities of project characteristics such as "technical risk" or "likelihood of commercial success." These are often impossible to quantify because of the uniqueness of innovation projects. The risk of using quantitative decision aids when valid quantitative data cannot be input is the old adage, "garbage in, garbage out." The danger is that formal selection methods may result in the wrong project selection as often as in the right one. Also generally ignored in the project selection literature as been the problem of validating a project-selection decision model. All models, decision or otherwise, need to be validated by experience or experiment before they should be used for serious business.

ROLE OF THE PROJECT MANAGER

The project manager is responsible for formulating and planning the project, staffing and running the project, and transferring project results to the customer. In doing this, the technical project manager needs to coordinate and integrate activities across multiple disciplinary and functional lines. In addition, all this must be performed within a finite budget and according to a firm schedule.

One of the problems with project management, from the perspective of the project manager, is that the project manager has a great deal of responsibility with only limited authority. The project manager does not control personnel positions, has a limited budget, must "sell" the project to upper management and to project personnel, and must also satisfy a customer. As Kerzner summarized the difficulties of the role:

> The project manager's job is not an easy one. [Project] managers may have increasing responsibility, but with very little authority.
>
> —(Kerzner, 1984, p. 10)

Richard Beltramini examined a set of product development projects and concluded that many factors contribute to the success of a project team. Among them are (1) how the team was organized, (2) how information was handled, (3) the timing of project and its performance, (4) training of the participants, and (5) resource availability (Beltramini, 1996).

Gary Gemmill and David Wilemon studied the kinds of frustration felt by technical project leaders, and found that the most frequent kinds were:

1. Trying to deal with apathy of team members
2. Wasting project time when the team goes back again over the same issues
3. Getting the team to confront difficult problems
4. Getting appropriate support for the project from the larger organization

(Gemmill and Wilemon, 1994)

MATRIX ORGANIZATION

As we have discussed, in organic organizational cultures where products are one-of-a-kind or customized for a particular customer, project management is the form of managing operations. In these organizations, authority structure is usually in the form of a matrix organization: functional divisions matrixed against product projects. Figure 12.5 illustrates such a matrix organization, with functional divisions listed vertically and product projects listed horizontally.

The functional divisions hire personnel by functional or technical back-

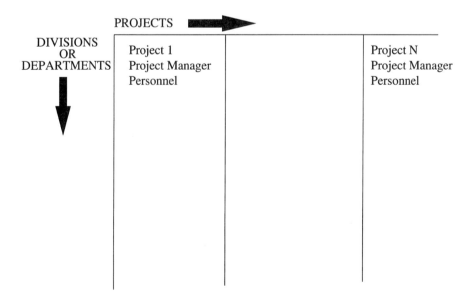

Figure 12.5. Matrix organization.

grounds and house them in a functional division of labor. These divisions have "personnel slots"; the managers of the divisions have authority to hire personnel, but not to pay them. In contrast, the projects have budget authority, but without the staff to perform the project. Accordingly, cooperation between division managers and project managers is necessary to staff and fund a project. This is why this form is called a matrix organization; personnel work for two managers: division and project.

The matrix form requires attention to staffing, monitoring, and supervision of personnel across two lines of authority. It is difficult to run, but necessary to control organic forms. Matrix organizations make it easier for the project manager role when they set up formal procedures that facilitate the form of project management in the organization. Some of the formal procedures that facilitate project management include:

- Full-time assignment of personnel on a given project
- Dual supervision of project personnel by project manager and functional manager
- Career advancement for project personnel and managers
- Organization-wide coordination and scheduling of all projects

In nonmatrix organizations, projects are fewer, and such organizations are less likely to have formal procedures to facilitate project management.

The advantages of a matrix organization is the ability of the organization to provide authority along more than one dimension of the organization (e.g., function and product, or geography and product).

INTEGRATING PROJECT MANAGEMENT INTO BUREAUCRATIC STRUCTURE

One of the classical problems with bureaucratic organizations in functional divisions is the coordination across functions. Since project management cuts across functional divisions, experience in a development project in a bureaucratic structure can help build cross-functional ties.

H. Kent Bowen, Kim B. Clark, Charles A. Holloway and Steven C. Wheelwright have argued that development projects can facilitate organizational renewal:

> A development project is a microcosm of the whole organization. A project team is made up of people from many areas of the company. A team's success is determined by the integrated outcome of everyone's work. . . . Because development projects typically are conducted under intense time and budget pressures, they usually magnify the strengths and weaknesses of a company. . . . Development projects provide a . . . real-time test of the systems, structures and values of the whole organization.
>
> —Bowen et al., 1994, p. 111)

Furthermore, they advised that companies should teach employees involved in a project that *how* the result is achieved is as important as the result itself. In this way, the goal of learning should be a primary goal of everyone in a project. Participating in this study were several companies: Chaparral Steel, Digital Equipment, Eastman Kodak, Ford, and Hewlett-Packard, and they provided examples of development projects for case studies. From the lessons they learned in the cases, they summarized several elements they thought were valuable factors in learning from development projects:

- Learning to appreciate the core capabilities of the company and its guiding vision
- Better understanding of the organization and its leadership
- Fostering senses of ownership and commitment by participants in the project
- Experiencing the pushing of the technical envelope of the company's products and processes
- Experiencing the development of prototypes of new product or processes
- Learning to integrate work across company functions
- Auditing development projects after completion to learn why some succeeded and others failed

SUMMARY

Project management in an organization may be focused on projects for either strategic or operational change. Technical projects are of three sorts: invention, design, or operations. A financial project chart graphs the projected costs and returns of the project and sales volume over time.

All technical projects need to be planned, implemented, and completed. In innovating new technology, project management techniques should be adapted to technical risks. The project manager is responsible for formulating and planning the project, staffing and running the project, and transferring project results to customer.

The project matrix organizational form requires attention to staffing, monitoring, and supervision of personnel across two lines of authority—function and project.

FOR REFLECTION

Define a project within an existing organization for a new product, process, or service in the organization. Create a project plan and project costs and benefits.

___13
HIGH-TECH
NEW VENTURES

CENTRAL CONCEPTS

1. Types and potentials of new ventures
2. Businesses opportunities in a new technology system
3. Key ingredients for high-tech new ventures
4. Timing of innovation
5. Dynamics of high-tech new ventures
6. Generic and proprietary technologies
7. Vision of a new enterprise

CASE STUDIES

Origin of the Personal Computer Industry
Jobs and Wozniak Start Apple
Rise and Fall of Osborne Computer
IBM's Failure to Dominate the Personal Computer Market

INTRODUCTION

We now shall return to the concept of the enterprise system, particularly focusing on the issue of starting new businesses based on technological innovation. We recall that industries based on new technology are called "high-tech" industries as long as rapid progress in their core technologies continues.

A high-tech new venture is the startup of a new company in an emerging industrial value-adding structure that supplies functionality around a new basic technology system. There are three major sources of risk in a high-tech new venture: (1) starting up a business, (2) the rate and directions of technical progress of the new basic technology, and (3) the rates and niches of market growth. But the importance of the new technology justifies the risks. For example, in studies of the success of new business ventures in the United States in the 1970s, high-tech new business ventures showed twice as high a survival rate than new business ventures not using high-tech (Vesper, 1980).

We will review the challenges of starting a new high-tech business venture of high-tech ventures, types, sources of ideas, and dynamics.

TYPES AND POTENTIALS OF HIGH-TECH NEW VENTURES

High-tech new ventures can market high-tech products or high-tech services. High-tech products can be new materials new components, new final products, or new process equipment. They can have large, mass markets or small, specialized markets. If the markets are large, the new venture has an initial possibility of growing into a large firm. If the markets are small and specialized, the new venture has only the possibility of becoming a niche-producer or being acquired by a larger firm.

High-tech services can be professional services, consulting, training, services requiring high-tech devices or procedures, or repair and maintenance services. New firms providing high-tech services generally require regional presence, and therefore usually have limited growth potential unless some form of national delivery of services can be constructed. Growing larger usually requires some form of partnership (e.g., large law firms or consulting firms) or a franchising method.

New high-tech software firms can provide either mass-market software or specialized customizing software. A new firm offering a mass-market software product has the potential of becoming large, but even a new firm offering specialized customizing software generally will also require regional location.

Thus, the potential sizes of new high-tech firms depend on the size of the potential market and/or regionalization of the market. Moreover, the larger the market, the greater the need for capital to grow to dominate it, and the greater the potential of large companies entering it with large capital resources.

Another factor influencing the potential size of new high-tech ventures is intellectual property. Basic patents that cannot be effectively invented around can be sources of ideas for new businesses and can provide protection from competition.

Case Study: Origin of the Personal Computer Industry

This case study illustrates how the devices and components of a basic technology offer business opportunities to found new companies and begin a new industry. The historical context was the origin of the personal computer industry. This began in the United States in the late 1970s as a kind of technological spinoff of advances in the computer and semiconductor chip industries.

In the 1970s, computers were being produced in three product lines: supercomputers, mainframes, and minicomputers. Semiconductor chip technology had begun that decade with the technical capability of forming thousands of transistors on a chip (then called middle-scale integration (MSI)) and, by 1975, had begun to be able to form tens of thousands of transistors on a chip (called large-scale integration (LSI)). This scale of improved capability was exploited by Frederick Faggin (an engineer at a then-small chip company called Intel) to design the key component of a computer system, the central processing unit (CPU), on a single semiconductor chip. The commercial motivation for Faggin's design was that Intel was producing chips for electronic computers, custom-designed for each customer. Faggin thought it more economical to produce one programmable chip, rather than many redesigns. He called it a microprocessor, and, in 1973, Intel produced it as the Intel 8008 chip. But Intel management did not then see the potential of the chip as the key component to build a personal computer, so Intel did not innovate the personal computer. The microprocessor was not a complete computer system, since it needed input and output devices and memory devices. But the core of the logical processes of the computer were there on a single chip.

At first, the computer industry ignored Intel's new chip, but other technophiles did not ignore it. They could see a new revolution on the horizon, a computer for everyone—the personal computer. Stephen B. Gray and other computer buffs had founded an Amateur Computer Society in 1966 (Gray, 1984). When Intel's chip was marketed, there were technical individuals just looking for the capability that the Intel chip provided. Nat Wadsworth, an engineer at DataCom in Danbury, Connecticut, suggested to his management that it use the chip to simplify the company's products, but they were not interested. Then Wadsworth talked to his "tech" friends: "Why don't we design a nice little computer and each build our own to use at home?" (Gray, 1984, p. 12)

Two friends joined Wadsworth, and they designed a small computer. Wadsworth decided to manufacture it, and in the summer of 1973 he quit his job and incorporated the Scelbi (Scientific, Electronic, Biological) Company. It was the first personal computer firm. Early the next year, an amateur radio magazine, QST, ran the first ad for a personal computer. Unfortunately, Wadsworth had a heart attack and was unable to vigorously build a new company. He sold about 200 computers, half assembled and half in kits. While in the hospital, he began writing a book, *Machine Language Programming for the 8008,* which sold well.

Meanwhile, another entrepreneur, H. Edward Roberts, was also building one of the first personal computers. He was then working at the Air Force base in Albuquerque, New Mexico. Earlier, he had formed a company with a partner, Forrest Mims, called MITS. In 1971, they manufactured electronic calculators, which at first made a profit. Roberts then bought out his partners in MITS. But by 1974, big firms had moved into the electronic calculator business, with better products and cheaper prices. MITS was then $200,000 in debt, and Roberts decided to make computers with the new Intel 8008 chip (Mims, 1984). At the same time, Arthur Salsberg, an editor at the magazine *Popular Electronics,* was looking for a computer project using the new Intel chip and learned of Roberts' project. Salsberg called Roberts and told him that if he could deliver an article in time for the January 1975 issue, it would be published. Roberts called the computer Altair, and it was featured on the front cover of Popular Electronics: "Project Breakthrough! World's First Minicomputer Kit to Rival Commercial Models." Salsberg titled his editorial in that January 1975 issue, "The Home Computer is Here!" (Mims, 1984, p. 27).

The publicity worked, and orders poured into the company. But the product was awkward and difficult to use. The first Altair in 1975 cost $429 in kit form and came without memory or interfaces. It had no keyboard and had to be programmed directly in binary by setting switches. But it was a computer. Roberts did use a standard bus (the S-100 bus), which became the first standard for the new personal computer industry. In 1977, Roberts sold MITS.

At the same time, because Intel management would not then envision the potential market of the microprocessor as the core of a personal computer, Frederick Faggin left Intel to start a new company, Zilog. Faggin produced an equivalent 8-bit microprocessor called the Z80 chip. A competitor, Chuck Peddle, at MOS Technology, designed another competing chip, the 6502. These two chips, the Zilog Z80 and the MOS 6502, were to become the basis of the expansion of the personal computer. Intel's chips were to become the key component only later in the 1980s.

In January 1976, Chuck Peddle showed the new MOS 6502 chip at a West Coast electronics trade show (WESCON). He sold it for $20 a chip, whereas Roberts had paid $100 a chip for the Intel 8008.

The two chips, the Mos 6502 and Zilog Z80, were the basis for the next generation of personal computers, produced by Apple, Commodore, and Atari.

Growth in the personal computer industry was based not only on the major devices of the microprocessor chip and personal computer, but also on products that completed and added to the computer system. One important innovation was the creation of a higher-level programming language that would be easy to learn on personal computers. The Altair had to be programmed directly in binary by setting switches. In 1975, Bill Gates and Paul Allen wrote a "Basic" language interpreter for the Altair (Gates, 1984). Earlier, "Basic" had been authored by a Dartmouth professor as an easier-to-learn programming language than Fortran, which had become the standard programming language for mainframe computers and was innovated by IBM researchers in the 1950s.

Marketing infrastructure to sell personal computers also had to be created. In 1975 in Los Angeles, Dick Heiser started the first retail computer store. At the same time, another entrepreneur, Paul Terrell, opened a retail store, the Byte Shop, in Mountain View, California (Ahl, 1984).

As these new companies demonstrated the new market, Intel, as well as IBM, became interested in the market.

By 1980, the major device, peripheral parts, application software, and retailing infrastructure had appeared for the new personal computer industry. A new market had begun.

BUSINESS OPPORTUNITIES IN A NEW BASIC TECHNOLOGY

The first business opportunity that occurs in any new basic technology is the production of a product embodying the technology as core. (For the personal computer industry, the major device to be produced was the personal computer.) However, devices are often needed to complete an application system. (For example, Roberts' Altair computer of 1975 it came without the necessary technical components of memory, interfaces, and keyboard. It also came without a operating system software to connect it all together to handle stored programs and external memory devices.)

Early business competition among firms manufacturing the major device will depend on the degree of technology system completeness of a major-device product and its performance and cost. Other business opportunities arise in adding other devices to the major device to complete an application system.

Other business opportunities arise from retailing opportunities and from application opportunities. (In the case of the personal computer, many new retail firms began until PCs became commodity-type products. Many new firms creating software applications also began.)

Case Study: Jobs and Wozniak Start Apple

One visitor at the 1976 WESCON who purchased a chip was Steve Wozniak, then a technician at Hewlett-Packard in Palo Alto, California. Wozniak constructed a personal computer, which he called the Apple I. He showed it in the spring of 1976 to a local amateur computer club, the Homebrew Computer Club. Again, it was only a partly complete computer—no keyboard, no case, no power supply, no external memory, no printer, and no software. Yet two friends of his in the club—Steve Jobs and Paul Terrell—were impressed. Jobs formed a company with Wozniak to produce the computer and Terrell ordered the first 50 units to sell in his Byte Shop (Ahl, 1984).

With Terrell's orders in hand and needing cash for production, Jobs sold his Volkswagen and Wozniak sold his two HP calculators. Jobs hired his sister and a student to assemble the units, and Wozniak stayed on at his job to keep an income. In 29 days, 50 computers were delivered to the Byte Shop at $666 each. A further 150 units were assembled and sold. By the end of that summer, Wozniak was designing a newer model. This was to have a keyboard, a power supply and plug-in slots for the S-100 bus (which was a means to add peripheral components to the computer).

But, for production expansion, much more capital was needed. Jobs and Wozniak went to a venture capitalist, who invested $91,000 of his own money and agreed to provide management support for the fledgling high-tech company. This was A. C. "Mike" Markkula, an electrical engineer who had worked for Fairchild and Intel when they were early high-tech growth companies producing the then-new semiconductor chips (Fairchild was one company at which the semiconductor chip was independently invented). As an early employee at Intel, which grew rapidly, Markkula's stock options there made him a millionaire, and he retired at age 34. Markkula had the business and management experience that Jobs and Wozniak did not have. Markkula then hired Mike Scott to run the new company, which they called Apple. They set out to make it a Fortune 500 company, and they succeeded.

There were several key ingredients in Apple's early success. First, the architecture of the Apple computer was published, so that others could design products to complete and improve the product. One of the first external memory products added for the Apple was the tape cassette and, later, the floppy disk drive. The disk drive was necessary to make the personal computer more than a mere toy. Next, word processing software was written for the personal computers. The first example was written by Michael Shrayer for the Altair as a "text editor," and Shrayer then expanded this into word processing capability, which he called the "Electric Pencil" (Shrayer, 1984). After Apple was introduced, software vendors began offering word processing software on the Apple as a useful application. Games were also programmed for the Apple.

A second major application for the Apple was the development of spreadsheet software. In 1978, Dan Bricklin was a student in the Harvard Business School. He and his friend Bob Frankston were doing an assignment in their consumer marketing class to analyze a Pepsi Challenge campaign. Bricklin had devised spreadsheet software to make it easier to do "what-if" analyses, as they needed to do for the Pepsi assignment. Bricklin projected the financial results of the Pepsi campaign for five years instead of the two years projected by all his other classmates, laboriously done by a hand calculator. The advantage and power of spreadsheet software to the business world was immediately apparent to Bricklin, Frankston and their professor. Bricklin and Frankston formed a new company and marketed the spreadsheet as VisiCalc at the National Computer Conference in New York in 1979 (Bricklin and Frankston, 1984). It ran on the Apple computer, and gained Apple a small foothold in the business market for the personal computers.

INGREDIENTS OF A HIGH-TECH NEW VENTURE

What does a new high-tech venture require? Five key ingredients are:

1. A new product, process, or service idea based on new technology—a high-tech product, process, or service
2. A focus on a potential market for the new high-tech product, process, or service
3. Organization of a new firm to produce and market the new product, process, service to its focused market
4. An initial venture team to launch the new high-tech firm, which includes technical, business, and marketing experience
5. Capital necessary to launch the new high-tech firm

We know that technology is not itself a product or service. Therefore, the idea of a new product or service that can embody the functionality a new technology provides is a separate idea from that of the technology itself. This is why innovation is a separate process from invention. Invention creates the idea of the technology, whereas innovation creates the idea of the product (or service or process) using the technology that can be sold to a customer.

Launching a new high-tech venture thus requires both understanding the new technology and its potential and understanding a customer who could use the technical functionality in a product, process, or service. This is the creative concept of a new venture—the synthesis of the new product, process, or service idea.

But it is not enough to have a vague idea of a customer. A precise market focus is necessary for a successful new venture. What is the class of customers? In what kind of application will the customer use the new product, process, or service? What are the performance requirements and features of such a new product, process, or service for it to be useful in such an application? What must be the price of the product, process, or service for it to be purchased by such a customer? How can that customer class be reached for advertising and distribution? What will be the nature of the sales effort required? What then must be the maximum cost of producing the product, process, or service? How fast can the market be grown?

With the product and market in focus, the next idea to make clear is the nature and size of the organization required to produce and sell the product to the market. What must be the beginning form and size of the organization to enter the market with the new product? When will competitors enter with a competing product, process, or service, and at what price? How fast must the organization expand to grow the firm to competitive advantage over competitors?

With the product, market, and organization in focus, the next idea to fix is the members of the initial management team for the new venture. The team must include persons experienced and skilled in (1) technology and product/process/service design, (2) production management, (3) marketing management, and (4) capital and financial management. Thus, the ideal new venture team should consist of four persons. One partner should handle product development, design, and testing. A second partner should handle production development, organization, and management. A third partner should handle marketing and sales organization and development and management. A fourth partner should handle capital development and management and development of financial control and reporting procedures. Of course, new teams are often begun with fewer than four persons, but each of the four areas of product, production, sales, and finances need to be covered, even initially, by one or more persons doing extra duty.

The fifth key ingredient in a new business venture is the requirement for capital. Capital provides the ability to gather and organize the resources necessary to begin or expand a productive operation. A successful productive operation will eventually create profits to return the capital, but the time delay between starting or expanding a productive operation and the later accumulation of profits to return the capital is the basic reason why capital is fundamental to business. Capital represents the "time" aspect of value-adding economic activities. Accordingly, all capital is evaluated not only on amount, but also on timing. In estimating the requirements of capital, the new venture team must carefully forecast its required amounts, when required, and the time line for returning investors' their capital.

TIMING OF INNOVATION AND FINANCIAL RISK

In launching a new venture, timing is critical. Integrating technology change and commercial application must occur at a time of market opportunity. There are several financial risks: (1) having the capital required to exploit the market opportunity, (2) timing the exploitation to recover the invested capital, and (3) generating sufficient working capital to meet competitive challenges.

The competitive advantages of a strategic technology occur within a "window of opportunity," followed by competitive challenges to this opportunity. When technology leaders introduce a new product that creates new markets, competitors will enter by focusing on the obviously weak features of the innovative product. Thus, for competitive advantage, technological innovation is never a one-time activity. It is a sequence of activities that follow-up in creating and sustaining a competitive advantage. The financial risks in timing technological innovation are many:

1. Having the required capital to perform the R&D necessary to design the new product
2. Having the required capital to produce the new product
3. Having the required capital to market the new product to the point of break-even on the investment
4. Having the capital to meet follow-up competitive challenges as the product and market evolve

Case Study: Rise and Fall of Osborne Computer

This case study illustrates the importance of timing on commercial success or failure during innovation. The historical context of this case study is still in the early days of the new personal computer industry. It was just after Apple's initial success and just before IBM entered the new market. In July 1981, Osborne Computer was started as a new company in the brand-new personal computer industry; and during its first two months of production, it booked its first $1 million in sales. In the second year of operation, its net revenues jumped to $100 million. Yet, only six months later, Osborne Computer was bankrupt (Osborne and Dvorak, 1984).

In 1980, Adam Osborne sold his computer-book publishing company to McGraw-Hill, and stayed on to work. But Osborne decided to create his own brand of personal computer. He left McGraw-Hill and started his own firm. He had seen a market opportunity in the new personal computer industry. At that time, personal computers were all being sold as components—computer, disk drive, monitor, printer, and software. Osborne decided to package them as a portable computer and to sell the package at a lower price than competitor's equivalent component sets.

Osborne hired Lee Felsentein, an engineer, to design the electronics for the computer. Osborne presented his ideas for the new venture to Jack Melchor, a venture capitalist. Melchor invested $40,000. The new computer was an instant success; by the end of the first year, sales soared to one-third of the then-leader, Apple. Osborne had created a new market niche in the personal computer industry.

Quickly, competitors entered the new market, focusing on the Osborne's visibly weak features—a small screen with a narrow column width. A new competitor, Kaypro, pounced on these weaknesses, introducing a similar model but with a 9-inch screen and full 80-character width. Kaypro's sales soared, and Osborne's sales collapsed.

Osborne had planned a public offering to generate additional capital. He intended it for the summer of 1982, but put it off until early 1983. But at that time, the brokerage firm decided not to make the offer because of the sales slump. Customers were either buying Kaypro or waiting for Osborne to offer a larger screen. In that summer, the sharp sales drop created a financial crisis. October, November, and December passed without significant sales, as did the following January, February, March, and April. The Osborne Company was hemorrhaging cash.

The company it could not get further bank loans, and its planned public offering could not be made with its lack of sales. Osborne made several attempts to raise new capital privately, but had no success. There was not enough cash to carry the company, and in September 1983, the Osborne Company declared insolvency under Chapter 11 of the bankruptcy law. All the millions of equity on paper, only a year before, had vanished!

Yet through the 1980s, the market for portable computers continued to grow, and it became the most rapidly growing segment of the PC market.

DYNAMICS OF NEW HIGH-TECH VENTURES

Timing, planning, and implementation are the critical factors in the success and failure of technological innovation, including new ventures. Timing is critical when the new product is introduced, when production is expanded, when the product is improved, when competition enters, when production cost is reduced, and when working capital is created. The dynamics of growth or death of new firms center around timing.

> *Product ideas, production expansion, marketing, controlling costs, creating working capital, meeting competition with improved products— the timing of these determines the dynamics of new ventures.*

Jay Forrester introduced analytical techniques for examining problems that arise from timing of activities in a firm, which he called "systems dynamics" of organizations (Forrester, 1961). Forrester then applied the technique to the startup and growth of new high-tech companies. Figure 13.1 shows the four most common patterns that Forrester found happen to new companies. The first curve, A, is an ideal pattern, for which all companies hope. Initial success is rapid and exponential in sales growth (as first occurred at Osborne); then, growth slows but continues to expand as the firm makes the transition from a small firm to a medium-sized firm to a large

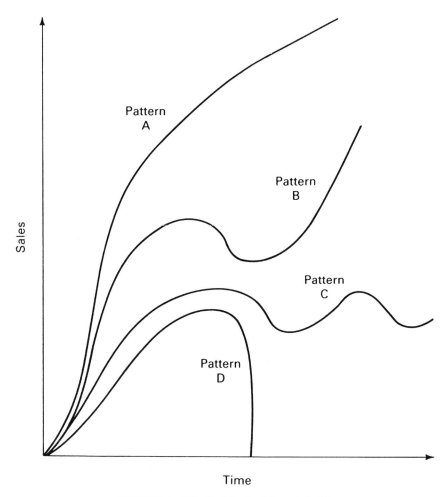

Figure 13.1. Growth patterns of new ventures.

firm. But curve A is relatively rare. The harsh facts are that few new high-tech firms ever become large firms.

Another successful pattern is curve B, in which a problem occurs soon after rapid expansion but the problems are solved and growth then resumes. Curve B is even rarer than curve A; usually, severe problems early in the growth of a new firm kill it, because the working capital of new firms is always thin and fragile (as happened at Osborne Computer).

For new companies encountering troubles, the most common pattern is curve D, when (like Osborne) the new firm's capital cannot sustain a period of losses.

But for high-tech new ventures, pattern C is frequent. Here, growth levels off as competitors enter the market, but the company successfully establishes a market niche for itself and continues on as a small-to-medium sized company, or is purchased by a larger firm.

What are the organizational factors that determine these dynamic patterns? Figure 13.2 shows a simplified version of Forrester's organizational systems dynamics modeling approach. In Forrester's modeling approach, it is important to distinguish organizational activities (boxes) from information states (circles). Organizational activities transform organizational inputs to outputs, while information states report on the results of transforming activities. In any productive organization, there will be squares identifying general transforming activities of: (1) parts inventory, (2) production, (3) production inventory, (4) sales, (5) product delivery, (6) market, and (7) revenue. In addition, important circles of information states will include (1) profits, (2) costs, (3) sales strategy, (4) product-order backlog, and (5) production strategy.

The dynamics of organizations occur from differences in timing of the activities in the different squares. For example, in an ideal case, sales strategy (3) will have accurately predicted the rates and growth of sales, upon which production strategy (5) will have planned and implemented adequate production capacity and scheduled production to satisfy sales projections and to meet projected product delivery schedules. Accurate timing will minimize parts inventory and product inventory and delays in delivering product to market. In turn, the market will pay promptly, allowing revenue to pay costs and show a profit.

The dynamics occur when all these activities are not properly timed. For example, if sales strategy overestimates actual sales, then production will

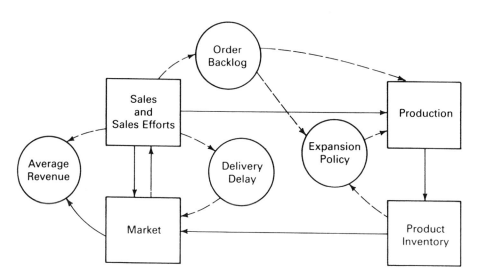

Figure 13.2. Systems dynamics of a new venture.

produce too many products, increasing costs from excess parts inventory, product inventory, and production capacity. If, on the other hand, sales strategy underestimates actual sales, the production capacity will be too small to meet demand, resulting in delivery delays which, in turn, result in lost sales to competitors—reducing revenue and lowering profits.

If the scheme shown in Figure 13.2 is expressed in a simulation model (organizational systems-dynamics model), then "what-if" games can be run to understand the sensitivity of product strategy, production strategy and sales strategy on revenues and profits. Then, the patterns of organizational growth as depicted in Figure 13.1 can be plotted for a given organization. Pattern D is, of course, the very sensitive pattern and always results from a cash flow crisis. When the costs rate overwhelms the revenue rate long enough to exhaust working capital and the ability to generate immediate new working capital, then such a firm will go bankrupt.

In the short-term, cash flow is always the critical variable for survival.

The critical delays in the activities that dramatically influence cash flow are:

• Revenue receipts that lag behind sales
• Sales efforts that lag behind sales projections
• Production schedules that significantly exceed product demand
• Delay in delivery of products sufficient to lose sales
• Significant delays and/or failure of market revenue to pay for sales

CRITICAL EVENTS OF HIGH-TECH VENTURES

Given these kinds of organizational dynamics of new ventures, there are nine critical activities in the startup and growth of new high-tech firms:

1. Acquisition of startup capital,
2. Development of new product or service
3. Establishment of production capabilities
4. Sales growth
5. Revenue growth
6. Production expansion
7. Meeting competitive challenges

8. Product improvement, production improvement, and product Diversification
9. Organizational and management development
10. Capital liquidity

Acquisition of Startup Capital

Capital is the resource necessary to begin and operate a productive organization with potential profitability until revenues can sustain the operation and provide profits. Startup capital is required for establishing a new organization and hiring initial staff, developing and designing the product or service, funding production capability and early production inventory, and funding initial sales efforts and early operations. Start-up capital is seldom sufficient for rapid growth; therefore, further capital requirements are usually necessary for commercial success.

Product/Service Development

A new firm is high-tech when its initial competitive advantage is in offering the technology advantage of new functionality, improved performance, or new features over existing products or services. Sometimes, new high-tech firms can be started with alternate high-tech production processes for existing types of products or services. But usually, a new high-tech product or service provides better competitive advantages with which to start new high-tech firms.

The next event is developing and designing the new product or service. This requires capital and will be a major cost in the startup venture capital. Ordinarily, development and design should be far along before startup capital can be attracted. However, development problems or design bugs that delay the introduction of a new high-tech product or service create serious problems in starting a new firm, because such delay also eats into initial capital. Moreover, if the delay is so long that competitors enter the market with a similar new high-tech product or service, then the advantage of having the first entry into the market is lost.

Establishment of Production Capabilities

The third event in a high-tech venture is to establish the capability to produce the new product or service. In the case of a physical product, parts and materials may be purchased or produced and the product assembled. The decision to purchase parts or materials or produce them in-house depends on whether others can produce them and whether or not there is a competitive advantage to in-house production. Establishing in-house production capability of parts and materials will require more initial capital than purchase, but is necessary when a given part or material is the innovative technology in the product.

However, the establishment of any new production capability will also create production problems: problems of quality, scheduling, and on-time delivery. Capital will also be required to "debug" any new production process.

Sales Growth

Initial sales and growth are the next critical event. The larger the initial sales and the faster sales growth, the less room there is for competitors to enter. An important factor influencing initial market size and growth is the application of the new product or service and its pricing. Another marketing problem is establishing a distribution system to reach customers. Distribution systems vary by type, accessibility, and cost to enter. Planning the appropriate distribution system for a new product or service, the investments to utilize it, and its cost influence on product/service pricing is important for the success of new ventures.

Generally, reaching industrial customers costs less then reaching general businesses or consumers. This is one of the reasons why a large fraction of successful new high-tech ventures are those in which industrial customers provide the initial market. These are usually industrial equipment suppliers or original equipment manufacturers selling to large manufacturing firms. This allows a new firm to get off to a fast start, but eventually limits the size of the firm and also places the firm in a position of vulnerability to backward integration by a customer. Moreover, a small firm with only a few industrial customers is very sensitive to cancellation of orders from any one of them.

In the general business and consumer markets, the distribution system infrastructure will usually consist of wholesale and retail networks. In these, access to the customer will depend on wholesaler and retailer willingness to handle the brand offered by a new firm. Establishing brand identity and customer recognition of the brand is then an important issue and a major barrier for a small new firm to overcome. Moreover, in some retailing systems, under-the-counter practices (such as buying shelf space and/or giving generous holiday gifts to purchasing agents) may also be barriers to overcome.

As a market grows, the long-term success of a new high-tech venture becomes increasingly important in gaining, and maintaining, access to national and international distribution systems.

Revenue Growth

Hopefully customers will pay their bills on time, but there is always some delay and bad debts. If a single large customer does not pay in time, a cash flow crisis may result.

Production Expansion

As the new market grows and sales are successful, production expansion must be planned and implemented in a timely manner or sales will be lost to competitors because of delivery delays. Production expansion will usually require a second round of capital raising, for the initial capital seldom provides enough for expansion.

Meeting Competitive Challenges

In a very few areas and rare cases, a patent on a new product or process is basic and inclusive enough to lock out all competitors for the duration of the patent. This is true in the drug industry and occasionally elsewhere. Most new high-tech ventures are launched with only partial protection from competition by patents, however, and competitors soon enter with "me-too" products. The me-too products or services are likely to be introduced with improved performance or features and/or at lower price. The entrance of competitors into the new market is the critical time for new ventures. They must, at that time, meet the competitive challenges or follow pattern C into a niche or pattern D into bankruptcy

Product Improvement and Diversification

A new firm must upgrade its first-generation products with new products to keep ahead of competition in product performance and features. It also must continually lower its cost of production to meet price challenges by competitors. And it must diversify its product into lines to decrease the risk that a single product problem will kill the firm. The round of capital raised for production expansion must also provide for product and production improvement.

Organizational and Management Development

As an organization grows in size to handle the growth in sales and production, it is important for the firm to develop organizational structures and culture and to train new management. This is an important transition, as the early entrepreneurial style of organization and openness and novelty of culture need to mature toward a stable but aggressive large organization. In a small firm, coordination is informal and planning casual. In a large firm, both coordination and planning need formalization.

Capital Liquidity

The final step for success in a new firm is to know when and how to create liquidity of capital assets and equity. One means is to go public; another is to sell the firm to a larger company. Liquidity of capital enables the founders of the firm and early employees to transform equity into wealth.

Case Study: IBM's Failure to Dominate the Personal Computer Market

This case study illustrates the need for proprietary technology for competitive advantage in the eventual success of new high-tech ventures. As in the previous case of the Osborne computer, the historical setting of this case is the growth of the new personal computer industry in the 1980s. IBM had been watching the market grow as the 1980s began, and, in 1981, it entered the market with an IBM PC. At the time, it had the superior technology of a new Intel chip with 16-bit word length, but it was not IBM's chip.

IBM's market share in the first year of its entry into the PC market jumped to 37 percent. IBM's brand name justified the new product to the business market as not just a "toy," but a true "computer." IBM alone had established the "industrial standard" for personal computers.

But, in the end, the story did not turn out happily for IBM. When IBM introduced its first personal computer, the product manager responsible for the introduction of the IBM PC failed to look at the long-term marketing consequence of his product design. Instead of drawing on IBM's technical capabilities and developing a proprietary operating system, he chose to license an operating system from Bill Gates' fledgling company, Microsoft. Also, instead of having IBM develop a proprietary microprocessor chip for the computer, he chose to purchase CPU chips from the semiconductor manufacturing firm Intel. Both choices turned out to be fatal for the long term position of IBM in the new PC market. Because IBM did not produce proprietary components of its product system, IBM introduced into the market a personal computer with only generic technology.

The new industrial standard that IBM helped establish was not based on IBM proprietary technology, but on Intel and Microsoft proprietary technology. IBM could not have conceived of a bigger favor to Intel and Microsoft. During the 1980s and 1990s, Intel and Microsoft prospered, while IBM's initially large market share in personal computers was eaten up by lower-cost "clone" PC manufacturers. IBM dropped from a high of

37 percent of the personal computer market in the mid 1980s to a low of 8 percent in the 1990s.

GENERIC AND PROPRIETARY TECHNOLOGY

Technology is implemented into products, production, service, and operations. Some of the technology implemented can be generic, publicly known, and some proprietary, known only by the company. Both generic and proprietary technologies are important to competitiveness, but in different ways. Generic technology is necessary but not sufficient to competitiveness, while proprietary technology is not necessary but sufficient.

What this means is that any competitor's technology must be up to the same standards as the best competitor's generic technology, for this is the publicly available standard of engineering. Products, production, and service that do not embody the most advanced generic technology are not globally competitive.

In contrast, proprietary technology consists of the additional details to make generic technology work completely in a product, production, service, or operation. Proprietary technology consists of the details that make the competitive differences between competitors who are all sharing the best generic technology available.

In products embodying technological innovation, product designs without proprietary technology invite imitators into the market.

STRATEGY FOR A NEW ENTERPRISE

Envisioning a new venture requires developing all the strategy necessary for an enterprise system. This requires two visionary components: (1) a vision of the transformation, and (2) a vision of how that transformation adds value to a customer.

Management's vision of the enterprise should provide answers to the following strategic questions:

- What businesses should we be in?
- What are the basic transformations of these businesses?
- What are the core technologies of these basic transformations?
- Who are the customers and how do we reach them?
- How do these transformations add value for the customer?
- What products or services provide that value?

- What competitive edge should we have over competitors in adding value?
- What core technologies provide the transformations in the product or service and in production?
- What technology strategies can provide competitive advantages?
- How do we staff, organize, and operate productively?
- What capital requirements do we have, and how do we obtain capital?
- What are our goals for return on investment, and how are they to be met?

In each value-adding transformation, subsets of strategy must also be addressed, such as:

- Practice of quality in the transformation—how good is good enough?
- Practice of customer service—how much and what kind of attention should be paid to customer requirements?
- Practice of profitability—what are minimally acceptable returns on investment?
- Practice of gross margins—how much overhead can we afford to have?
- Practice of rewards—how much of the profits should be divided among management, labor, and shareholders?
- Practice of control—what tools should we use to control corporate performance?
- Practice of innovation—do we compete as a technology leader or follower?

SUMMARY

A high-tech new venture is the founding of a new business based on a new high-tech product or service. The major device in a new technology system offers the first business opportunities to start new high-tech firms. Early business competition among firms manufacturing the major device will depend on the degree of technology-system completeness of a major-device product and its performance and cost. Other aspects of a new basic technology system will also provide business opportunities: components and connections, peripheral devices, control and software, services, retailing, training, maintenance, and repair.

The potential sizes of new high-tech firms depend on the size of the potential market, regionalization of the market, and/or intellectual property protection. Critical events must be met along the way in a new venture for commercial success. The timing of product development, production expansion, marketing, controlling costs, creating working capital, and meeting competition with improved products all determine the dynamics of new ventures.

FOR REFLECTION

Identify some new firms that went public in the last five years. Find their initial offering prospectuses and trace their stock prices since then. Have any encountered problems? What were the problems, and why did they occur?

___14

ENTREPRENEURSHIP

CENTRAL CONCEPTS

1. Entrepreneur versus bureaucrat
2. Vision and leadership
3. Sources of high-tech entrepreneurs
4. New venture teams
5. Psychology and sociology of entrepreneurship
6. Venture capital
7. Business plan for a new venture

CASE STUDIES

Al Shugart and Disk Drives
Starting Cymer Inc.

INTRODUCTION

Beginning a new high-tech business venture requires a style of management that is entrepreneurial, a style that we earlier noted is appropriate to the starting and building of new businesses. Accordingly, we need to review the concept of entrepreneurship for new high-tech business ventures. In the many studies on entrepreneurship, three themes have emerged:

1. *Adventure:* The entrepreneur is a kind of business hero, creating new firms and causing economic change. An entrepreneur has the qualities to be admired as the hero or heroine of an adventure: vision, initiative, daring, courage, and commitment.
2. *Organizational transition:* The entrepreneur makes a good manager for starting firms, but is not necessarily good for building organizations and growing firms. Because the entrepreneur plays many roles in the early stages of a firm, the transition to building an effective large organization is often difficult for the entrepreneur.
3. *Organizational dilemma:* To what extent and how can entrepreneurship be encouraged within a large, stable firm in order to manage change? Are bureaucratic management and entrepreneurship styles eternally at odds with one another in the same large organization?

Case Study: Al Shugart and Disk Drives

This case study illustrates the opportunities and hazards that entrepreneurs encounter. The historical setting is the decades of 1960 to 1990, when the hard disk drive was innovated for computers.

The day after he finished college in 1951, Al Shugart joined IBM. IBM, in the early 1950s, was just beginning to make electronic computers; a key component was permanent memory. Hard disk drive memory storage was innovated in 1961, and Shugart led IBM's development of the technology. But in 1969, Shugart resigned from IBM:

> [W]eary of Big Blue's bureaucracy, he [Shugart] left for Memorex Corp., taking along some 200 engineers.
> —(Burrows, 1996, p. 18)

Only a few years later, in 1972, Memorex had financial troubles, and Shugart left, again with loyal followers, and founded Shugart Associates. Shugart had the innovative idea of pioneering the floppy disk. But in 1974, the new company lagged in its product development for the floppy disk, and Shugart's venture backers ejected him:

> Nearly broke, he [Shugart] moved to Santa Cruz, opened a bar with some friends, and bought a salmon-fishing boat. "I had a tough time meeting my Porsche payments," he deadpans.
> —(Burrows, 1996, p. 18)

Four years later, in 1978, a old colleague, Finis Conner, went to visit Shugart in Santa Cruz. He suggested making small hard drives for the new personal computers just beginning as a market. They started Seagate, and the company rode the growth of personal computers, growing to $344 million in sales by 1984. But then, price wars began, as many producers marketed small hard disk drives. Shugart and Conner argued over production strategy. Shugart insisted on making drives from scratch, while Conner argued that buying parts was less risky because it tied up less capital.

Dissatisfied, Conner left and launched Conner Peripherals, with capital backing from Compaq Computer. In its first year, Conner made $133 million in sales, since purchasing components is a fast way to build production. In the short term, Conner was right, but the long term was not over.

Because of the intense price competition in the personal computer hard disk drive market, Shugart switched to the mainframe disk drive market (the one in which, a decade earlier, Memorex had stumbled against IBM). Shugart bought Control Data Corp's Imprimis Technology, which produced mainframe disk drives. Shugart survived by having two product lines. Next, as the PC market demand grew rapidly, key components became scarce, raising component prices. Now it was Conner's company that was in trouble. Along with other PC hard disk drive manufacturers who only assembled purchased components, Conner had no control over manufacturing costs. He and other assemblers lost money. In 1993, only Seagate made a profit in PC hard disk drives.

By 1995, Conner was in deep enough trouble that he needed to sell his company. His old colleague, Shugart, offered to buy it, giving Conner a job at Seagate. Shugart's make-versus-buy production strategy proved to be right for the long run. In 1996, Seagate projected revenues of $9.2 billion, up 104% from the previous year, and profits were projected to increase 130%, to $600 million.

Shugart's entrepreneurial style included correct technical vision and correct long-term cost strategy. He also had the ability to generate staff loyalty:

> "He's the most up-front guy I've ever worked with," says Vice President Stephen J. Luczo, whom Shugart lured from Bear, Stearns & Co. to build the software business.
>
> —(Burrows, 1996, p. 73)

ENTREPRENEUR VERSUS BUREAUCRAT

We recall that the customary usage of the term "entrepreneur," as compared to the term, "bureaucratic manager," is to distinguish two different types of leadership in starting economic activities, as opposed to running ongoing economic activities. Entrepreneurs start new firms, while managers run ongoing firms. There are important differences in the optimal leadership styles for starting versus running. Often, good entrepreneurs make poor managers of ongoing operations; and often, good managers make poor entrepreneurs when presented with the challenge of dealing with substantial change.

Howard Stevenson and David Gumpert compared managers along two dimensions: (1) desire for future change and (2) perceived ability to create change (Stevenson and Gumpert, 1985). They claimed that the entrepreneur is the kind of manager who desires future change and perceives the ability to create such change and that the entrepreneur asks the following kinds of questions:

- Where is the opportunity?
- How do I exploit it?
- What resources do I need?
- How do I gain capital control over them?
- What structure is best to exploit the opportunity?

In contrast, Stevenson and Gumpert noted that managers of ongoing operations are more concerned with stability and efficiency than with change, and, accordingly, ask a different set of questions:

- How can I improve the efficiencies of my operations?
- What opportunity is thus appropriate?
- What resources do I control?
- How can I minimize the impact of others on my ability to perform?
- What structure determines our organization's relationship to its market?

We recall that the term "bureaucrat" need not connote ineptness, but may connote "rationality." There can be good or bad managers, whether entrepreneurial or bureaucratic. We will use the concept of the "bureaucratic manager" to emphasize a managerial concern with "rational order" for an organization—the "good" bureaucrat. The tools of rational order are formalized procedures that routinize organizational efficiency. When organizational change is needed, then an entrepreneurial management style is useful; but when organizational stasis for rational efficiency is needed, then a bureaucratic management style is useful.

VISION AND LEADERSHIP

Whether managers are entrepreneurial or bureaucratic, an essential leadership quality of all management is correct vision. Leadership in management can span many aspects of management activity; but there are three central and critical attributes of organizations that only the top leadership can provide. These are vision, resources, and delegation. Vision establishes the direction of the organizational activities, resources the base for activities, and delegation the team to carry out the activities. In all organizations, these three attributes are the primary responsibility of the leader. If the leader does not fulfill these responsibilities, the organization will be either directionless, undercapitalized, or lacking an effective team.

For managing technology, two visions are essentially important. The first is a vision of the enterprise system, and the second is a vision of the technologies for the enterprise system. The vision of the enterprise system needs to define the nature of the business, customers, and competitive strategy. The technological vision needs to identify and anticipate technological change for the technologies involved in the major subsystems of the enterprise system. Vision is an ephemeral quality, but it distinguishes quality leadership from mediocre leadership—whether entrepreneural or bureaucratic.

PSYCHOLOGY OF THE ENTREPRENEUR

Many have studied the psychology of entrepreneurs, hoping to learn why some people are more likely than others to become successful entrepreneurs. Researchers have listed several attitudes and values they found typical of the successful entrepreneur. This list includes such attitudes as:

- A desire to dominate and surpass
- A need for achievement
- A desire to take personal responsibility for decisions
- A preference for decisions with risk
- An interest in correct results from decisions
- A tendency to plan ahead
- A desire to be one's own boss (Vesper, 1980, p. 9)

All these attributes are positive, and reflective of the "heroic" myth of the entrepreneur.

But there may also be negative attributes. For example, the entrepreneur may also be self-centered, temperamental, authoritarian, and unable to stand grooming a competent successor. One can see both positive and negative attitudes in biographies of entrepreneurial people.

SOURCES OF HIGH-TECH ENTREPRENEURS

High-tech entrepreneurs usually come from larger organizations in which they first gain experience, training, or research capability. In fact, most successful high-tech entrepreneurs start firms with innovative product or service ideas for the markets of the large firms in which they earlier served.

An important experience that entrepreneurs bring to a new high-tech venture is a detailed and experienced understanding of a market. Second is an innovative idea for a new product or service for such a market. Third, entrepreneurs can bring to a new venture prior knowledge of production. A fourth important experience is in finance, budgeting, and control. These kinds of experiences come from entrepreneurs' having prior experience in the marketing, manufacturing, engineering, research, or financial control functions of a large firm.

Sometimes, entrepreneurs try to get their superiors interested in the new idea; failing that, they leave and begin their own companies. Sometimes, the innovative ideas originate in a university or government research lab, and researchers then start a new firm.

VENTURE TEAM

New ventures can be started by single entrepreneurs or by small teams of entrepreneurs; however, a good initial entrepreneurial team is a strong factor in launching a new high-tech business. For the reasons of market, product, production, and finance, a successful new high-tech venture is often a team effort:

1. One member of the team has prior market experience.
2. One member has research/technical experience.
3. One member has production experience.
4. One member has financial control experience.

In many new ventures, a single entrepreneur may play several or all of these roles. However, technical, production, marketing, and finance all require quite different skills and talents. Moreover, if a new venture grows rapidly, several persons will be required to control operations. One should try to assemble the new venture team as early as possible in order to increase the chances of a successful venture.

SOCIOLOGY OF ENTREPRENEURIAL SYSTEMS

Another approach to understanding the entrepreneur has been not psychological, but sociological. For example, James Quinn viewed the entrepreneur as a kind of role encouraged by an "individual entrepreneurial system," which is to say, a capitalistic system that encourages and supports individual initiative. Quinn identified several characteristics of an entrepreneurial system that encourage technological innovation:

- Fanaticism and commitment
- Chaos acceptance

- Low early costs
- No detailed controls
- Incentives and risks
- Long time horizons
- Flexible financial support
- Multiple competing approaches
- Need orientation (Quinn, 1985)

Quinn viewed the single-minded dedication of the entrepreneur as a kind of fanaticism, and stated that an economic or organizational system should tolerate the kind of ruthless, dedicated purpose required of an entrepreneur. The contexts of such single-mindeness will appear chaotic and disorganized, because the entrepreneur is fixed on the goal and will use whatever means or expediency that proceeds toward that goal. The economic/organizational system should be able to tolerate this apparent chaos, which includes little detailed control in the early phase of a new venture. The origin of new ventures operates in an opportunistic, cost-cutting, short-cutting way to a single-minded, clear-cut goal. The system should provide appropriate rewards for risk-taking in entrepreneurship. Rewards must be structured for long-term horizons, since it takes time for anything really new to become a success. Because of the experimentation and learning that go into new ventures, it is important for the system to provide flexibility in financing from many sources and for multiple and competing approaches. In the early days of any radical innovation, new ways are being tried out, and only down the line will an optimal configuration emerge for a standard design of a new technology. Need orientation should always be the goal of entrepreneurship, since satisfying a customer's need is the sure way to economic success. Systems that encourage the fulfillment of needs of a marketplace stimulate innovation that endures and is economically important.

One can see the logic in this list. Individual entrepreneurs need to be encouraged to show the commitment to start new activities and push them to success. An economic system should be loose enough to tolerate chaotic activity and allow for low costs of startup, and so on. A very important institutional factor for new high-tech ventures is sources of venture capital.

Not all systems encourage entrepreneurship. There can be many organizational barriers, such as:

- Perceived irrelevance of an innovation
- Punishment for taking a risk and failing
- Rewards only for short-term achievements
- A culture of "not-invented-here"
- Stodgy and unchangeable conventions

For innovation to be encouraged in a large firm, a company culture toward innovation is very important. Particularly important are the criteria about the "perceived irrelevance" of a new business venture and about the attitude of the firm toward risk taking. One of the striking facts about innovation in large firms is how little innovation occurs from very large research efforts in some firms. This is the result of the firm's conventions about its "enterprise system."

For example, one of the largest and most technically advanced corporate research laboratories was the General Electric Corporate Research Center in Schenectady, New York. In the 1980s, the average number of patents obtained per year was in the hundreds, but a new business in GE begun from a research idea was a relatively rare occurrence. In GE, the strategy of the CEO is for any business in GE to be in a large market and first or second in market dominance. The requirement for very large markets discourages risk taking to start new businesses. In contrast, 3M, also a large company, encourages innovation, annually launching many new products and some new businesses each year. 3M's corporate goal is to obtain 25% of its sales from new products every five years. Both GE and 3M are very successful firms in the 1990s, but have markedly different company cultures toward innovation.

Case Study: Starting Cymer Inc.

This case study illustrates the start of a new high-tech company where the entrepreneur's technical vision correctly met an anticipated market need. The historical setting is the mid-1990s, when the world's IC semiconductor chip industry continued its pace of rapid technological progress in increasing transistor density on a chip by reducing feature size. We recall that progress in chips proceeded exponentially through a series of next-generation technologies:

> [O]ver the past 30 years . . . chip makers doubled the number of transistors on a microchip every 18 months. For example, Intel's Pentium, vintage 1993, had 3.2 million transistors; its Pentium Pro, released about two years later, uses 5.5 million transistors.
>
> —(McHugh, 1997)

Chips are produced by projecting circuit diagrams onto photographic coatings on silicon wafers and then etching the photographic pattern into morphological features. As feature sizes of transistors approach the wavelength of the projecting light, then that light can no longer produce accurate features. (This is the result of the wave nature of light, whose diffraction patterns around objects ultimately limits the resolution of the images at a given wavelength of light.)

In the 1960s, chip makers used visible light. By the mid-1990s, chip makers used the shorter wavelengths of invisible ultraviolet light, produced by hot mercury gas, getting chip feature size down to 0.35 micron.

Next, chip makers hoped to jump from mercury light down to X rays, which have much smaller wavelengths, but by 1997, X rays still turned out to be too costly and difficult to use for chip making. The new light source for chip photolithography was excimer laser light, at a 0.25 micron wavelength. In 1997, these lasers were produced by a relatively new firm called Cymer Inc. (McHugh, 1997).

Back in 1985, two recent Ph.D.s from the University of California at San Diego, Robert Akins and Richard Sandstrom, were considering their future. Akins' thesis research was in optical information processing, which uses lasers, and Sandstrom's thesis research was in laser physics. They were working for a defense contractor, HLX Inc., on esoteric projects that used lasers, such as laser-induced nuclear fusion and satellite-to-submarine laser communications. Since they both wanted to make real money, they speculated about opening a business, and decided to use their special talents. They founded Cymer Laser Technologies in 1986 to build excimer lasers.

Excimer lasers could produce laser light by using a mixture of gases, krypton and fluorine, pumped into a two-foot aluminum tube, which is then "zapped" by a 12,000-volt charge across two electrodes inside the tube. The voltage creates a charge in the gases for 75 billionths of a second that excites the krypton atoms to temporarily couple with the fluorine atoms. This forms a krypton fluorine molecule as an excited dimer (hence the term "excimer"), but the unstable molecule breaks apart as soon as the voltage drops. In doing so, it releases a burst of deep-ultraviolet laser light of 0.25 micron wavelength.

In 1986, Atkins and Sandstrom began building the prototypes of this excimer laser in the research labs of the university (their new company owed the University of California $250,000 for the use of its facilities). Cymer also succeeding in winning research funding from the U.S. government's defense agencies to continue the development work for the laser. The tough technical problem was getting the laser to run dependably for months at a time, while handling the 1000 jolts per second of 12,000-volt charges. During this time of product development, Akins and Sandstrom took out second mortgages on their homes to keep Cymer running.

In 1988, Cymer received its first outside investment from a venture capitalist, Richard Abraham. Abraham had been a researcher and factory manager at Fairchild Semiconductor, Motorola, and Texas Instruments. He understood the importance of the deep-ultraviolet laser to the future production needs of the IC chip industry. Abraham's condition for investing was that "Cymer had to focus entirely on semiconductor applications for its lasers" (McHugh, 1997, p. 156). Later that year, further investments came from Canon and Nikon, after teams of their scientists and executives made visits to Cymer. Canon and Nikon manufactured the photo equipment, called "steppers," that semiconductor manufacturers used to step the photo of a circuit over a silicon wafer, and they would be customers for Cymer's laser. They bought 6% of Cymer.

Finally, in about 1995, semiconductor manufacturers realized that the mercury light technology had hit its natural limit and began the switch to excimer laser light for production of chips:

> Demand from chip makers forced the stepper companies [such as Canon and Nikon] right into high-volume buying soon after Cymer launched its $450,000 excimer laser model.
>
> —(McHugh, 1997, p. 156)

Cymer's sales exploded, from $18 million in 1995 to $65 million in 1996. In September 1996, Cymer sold stock to the public at $9.50 a share and, in December, at $42 a share. It raised $80 million in the two offerings. In February 1997, Cymer stock traded at $50. Akin's stake was 2 percent and worth $2 million.

VENTURE CAPITAL

Venture capital is a form of capital invested in new businesses. There are several stages of funding needed for new businesses: seed capital for startup, production capital for building production and distribution capability, and expansion capital for expanding production and distribution capability. Venture capital is a higher-risk investment in businesses than later forms of business investment, because few business assets exist at the time of venture funding; correspondingly, returns on successful venture capital investments can be very high because prices of equity shares are much lower before a firm proves itself.

Sources of venture capital include investments by the entrepreneurs themselves, technically-oriented individuals who have become wealthy and are sometimes called "investment angels" (such as Abraham), venture capital firms, venture capital funds, and venture capital investments by large firms looking for new technology (such as Canon and Nikon). Guides to sources of venture capital in the United States have been published periodically (e.g., Pratt and Morris, 1984). In launching a new high-tech business, it is usually difficult to attract venture capital investment until the product prototype is well under development.

In the Cymer example, the entrepreneurs had to use the university's facilities, government contracts, and second mortgages on their homes to get started. When they had a demonstration prototype to show, they then were able to attract startup capital, first from an individual and then from user firms, Canon and Nikon. The company went public just after sales of its product exploded, with the second offering generating more capital than the first offering.

The general criteria on which venture firms evaluate business proposals are product differentiation; market attractiveness; managerial capability; environmental, safety, or product liability; and cash-out potential.

However, there are differences among venture firms in where they will invest their capital. Edward B. Roberts pointed out that venture capital firms differ by the technologies in which they prefer to invest and also by preference for the stage of venture investment. Few venture capital firms will provide seed money for startups, with most firms preferring later-round funding to reduce their risks (Roberts, 1991).

In the second half of the twentieth century, the venture capital industry in the United States has played a major rule in growing new industry. For example, in 1992, venture capital investment was $4 billion, and was over $7 billion in 1995. Seventy percent of the 1995 investments were in information technology. Earlier venture investments had created large U.S. firms such as Sun Microsystems, Intel, and Microsoft.

Venture capital as a distinct industry slowly spread globally. In 1995, Europe had about 500 firms that invested a total of $9 billion in 20,000 companies. More than half of these investments, however, were not to start new companies, but to finance changes of ownership, mostly as management buyouts. In Asia, venture capital in the 1990s was still mostly corporate investment by giant conglomerates and family-run businesses (*Economist,* 1997).

THE BUSINESS PLAN

All new businesses need a business plan, and venture capitalists require one before considering an investment. The business plan expresses in detail the entrepreneurial vision for a new enterprise system. The purposes of the business plan are:

1. To chart the course and identify the resources needed for the new venture, and
2. To attract venture capital.

A business plan should cover the following topics:

1. Executive summary
2. Enterprise strategy and technology strategy
3. Product/service strategy
4. Competition and benchmarking
5. Manufacturing and marketing strategy

6. Management team and organization
7. Financial plan

Since the new business venture does not yet exist and, therefore, the plan cannot be tested against past performance, identifying assumptions in the plan is, clearly, very important—both for credibility and for estimating business risk.

Executive Summary

This is a one-page summary of the highlights of the business plan, written last and placed first. Its purpose is to generate sufficient interest by a potential investor to read the whole plan.

Enterprise and Technology Strategy

The enterprise strategy should express the concept of the business, its need, where it begins, and to where it should grow. The concept of the business should begin with the functional capability to customers, who they are, and their application needs. Then, the concept should identify what kind of product or service the new business will provide for that customer, and how much it will provide value to the customer. It is important to be very explicit about this assumption of adding value to the customer, for the pricing of the product or service depends on how much value it provides for the customer. Translating "value-addedness" into "price" is one of the most critical assumptions an entrepreneur will make in a business plan.

The technology strategy should identify and discuss the technological innovation that provides the value-addedness of the new business opportunity. When the new technology is in a product or service, then the discussion should indicate how that will change the customer's applications or create new applications. When the technological innovation is in production or organization for an already existing and standard product or service, then the value-addedness must translate into significant cost reduction and production quality improvement for the technological innovation to provide an entering competitive edge for a new business going up against existing competitors.

Product/Service Strategy

The next section should describe the concept of the new product or service that the new business will initially produce and market. The technical specifications of the product or service should be detailed, along with the current state of the development and design of the product. If the product is still in development, then a development schedule should be given, and technical risks in the development identified and described. Careful attention should be paid to identifying the technical risks and schedule for two reasons. The first is to make sure that sufficient capital is raised to carry through the development and begin manufacturing. The second reason is to protect the entrepreneur, by due diligence, from potential lawsuits by investors if development fails or falls behind schedule.

It is also important to identify not only the initial product or service in the business plan, but also a planned family of products and product lines and services that the business will evolve. It is rare that a single product will be sufficient to build a successful company. A product family and product lines are usually necessary for long-term commercial success.

Competition and Benchmarking

The competitive strategy should identify the way the new business intends to compete and what its planned competitive advantages are. It is important to benchmark competing products or products for which the new product or service may substitute. An important feature of such benchmarking is specifying technical performance and features of competing products and their prices. The plan should show the rate of anticipated technology diffusion of the new product or service into the market, and critical assumptions that facilitate or hinder that diffusion.

Manufacturing and Marketing Strategy

The business plan must also envision how the new product or service will be produced and the capital required to establish production. Production planning will require judgments about what parts and materials to produce and how to fabricate or assemble the product or service. The tradeoff judgments here are capital and learning costs of establishing production versus loss of control over proprietary knowledge and costs through purchasing. The advantages of producing in-house are that costs and quality can controlled and proprietary technology used in design and/or manufacturing, but this comes at capital costs. The advantage of outsourcing and purchasing parts and even fabrication is to lower capital requirements even though this cannot provide any proprietory competitive advantages over competitors that can source the same parts and materials.

It is also important to estimate the capital required to expand production. For a new venture that is quickly successful, the most common way for an entrepreneur to continue to dilute equity is to need second and third rounds of investment to expand production.

The marketing strategy needs to identify the potential customers for the new product or service and the applications context in which these customers will use the new product or service. The marketing strategy should also identify the customer requirements for the product or service and the price bracket in which the customer may pay for the product or service. The marketing strategy should identify the distribution channels for getting the product to the customer, and the costs and problems in setting up or entering these distribution channels. The marketing strategy should also plan the sales force, how the salespeople are organized and rewarded. The marketing strategy should identify efficient and effective means of advertising and distributing information about the new product or service to potential customers.

Management Team and Organization

The experience and credentials of the management team for the new business should be described. This is very important, because experiences of successful venture capitalists have emphasized that what investors are basically investing in is management. The organization of the business and operating procedures should be planned.

Financial Plan

The financial plan should begin with a sales projection and planned growth and penetration of the market over the first five years of operation. For the sales projections, the financial plan should forecast income, expenditures, and profits. In addition, working capital and balance sheets should be constructed for each year. Additional needs of further financing should be identified and discussed.

The financial plan should show projected return to investors as increase in equity. Finally, the financial plan should have a "cash-out" plan for investors and entrepreneurs to gain liquidity.

An interesting example of training to encourage entrepreneurship and the writing of new business plans is an annual student competition in business plans held by the Massachusetts Institute of Technology. In 1996–97, the MIT Enterprise Competition offered a prize of $50,000 to the student submitting the best business plan for a new company. Information about the competition was posted on the World Wide Web (http://web.mit.edu/50k/www). The competition was limited to MIT students, but nonstudents could team with an MIT student to submit a business plan. The competition announcement summarized the purposes of a business plan and its audiences:

- A business plan is a document that conceptualizes the totality of a significant business opportunity for a new venture;
- Presents the organizational building process to pursue and realize this opportunity;
- Identifies the resources needed;
- Exposes the risks and rewards expected; and
- Proposes specific action for the parties addressed.

Audiences for a business plan include the founding team, potential investors and employees, customers, and suppliers or regulatory bodies (MIT 50K, 1996).

SUMMARY

High-tech entrepreneurs usually come from larger organizations in which they first gain experience, training, or research capability. The psychology of entrepreneurs includes heroic qualities such as ambition, risk taking, independence, problem solving, and the like—all qualities useful in the chaotic conditions of starting a new business. However, these same qualities can later have negative impacts when trying to build a large organization, as they appear authoritarian and arbitrary to subordinates. Studies have also stressed the sociological conditions that foster entrepreneurship.

Venture capital is important for a new high-tech business: seed capital for startup, production capital for building production and distribution capability, and expansion capital for expanding production and distribution capability. Venture capital is a high-risk, but potentially high-payoff, kind of investment. All new businesses need a business plan to chart the course and identify the resources needed for the new venture and to attract venture capital.

FOR REFLECTION

Identify a new technology and imagine a new product, process, or service based on the new technology. Write a business plan for a new company. Would you encourage your relatives or friends to invest in the new venture? Why or why not?

____15
ENGINEERING FUNCTION

CENTRAL CONCEPTS

1. Engineering function
2. Product model lifetimes
3. Engineering logic
4. Engineering profession
5. Engineering disciplines
6. Managing the engineering function

CASE STUDIES

The Minicomputer Word Processor in the 1970s
Vannevar Bush, an Engineer of the Twentieth Century

INTRODUCTION

We now turn to the engineering function of productive organizations. The engineering function specializes in creating and implementing the technical bases of the enterprise. Business operations are divided into functional areas in order to delegate authority and assign responsibility for operations. The traditional functions of a business include production, marketing, finance, administration, engineering, and research.

We will review the engineering function of a business and its integration with the other business functions. We will address the following issues:

- What is the engineering function?
- Who are engineers?
- How should engineering be organized and integrated in the enterprise system?
- What are the possible career paths of engineers in the firm?

ENGINEERING FUNCTION

An engineering function is necessary in a firm for:

- Technological innovation
- Product and production design
- Provision of technical services

Successful technological innovation requires both new technology and market focus; this integration is the responsibility of engineering. All products have finite lifetimes and, to remain competitive, have to be periodically redesigned by engineering. Production to remain competitive needs to be continually improved. Technical services, both internal and external, must be provided by engineering.

PRODUCT MODEL LIFETIMES

We recall that product lines had finite lifetimes when a core technology of a product line became obsolete. But even within a product line lifetime, there are shorter lifetimes for a product, a product model lifetime.

A "product model" is a product designed for a market niche and a price/performance target. Product models exhibit the same functionality but vary in performance, features, fashion, and price. There are five principal reasons why products have finite lifetimes:

1. Technical performance obsolescence
2. Technical feature obsolescence
3. Cost obsolescence
4. Safety
5. Fashion changes

Performance or feature obsolescence in a product occurs when its performance and/or features are less than a competitor's model. Cost obsolescence occurs in a product when the same performance can be obtained from a competing product at a lower price. Safety obsolescence occurs when a competing product offers similar performance and price with improved safety of operation (or when government regulations require safer features or operation in a product line). Finally, when technology, costs, and safety features are relatively stable, products can still become obsolete due to fashion changes (such as in clothing). Fashion obsolescence occurs in a product in which product competition is not differentiable in performance and price but is differentiable in lifestyle.

Case Study: The Minicomputer Word Processor in the 1970s

This case study illustrates innovative engineering design. This project launched a new model of minicomputer-based word processors. Initially, it was a big success, but soon failed after its product line was rendered obsolete by personal computers. The historical setting is in the late 1970s when, we recall, semiconducting integrated circuit (IC) chip technologies began to be capable of containing, on a single chip, the central part of the computer (its central processing unit, CPU); this was called a "microprocessor." An Wang pioneered the early dedicated applications of the minicomputer to word processing.

An Wang had been a professor of electrical engineering at the Massachusetts Institute of Technology in the early 1950s. He invented the first magnetic core type of memory circuit for the then-new electronic computer technology. This important invention was purchased by IBM for a $1 million. Wang resigned from MIT and started his own electronics firm with the capital.

At first, Wang designed and built electronic calculators, but larger firms entered the field, driving his small operation out of business. Wang needed a new product, and introduced a minicomputer-based word processor in 1971. Then, larger companies also introduced minicomputer-based word processors. On September 28, 1975, An Wang held a strategy meeting with three of his employees. He said he was worried about their current word processor facing new competition from IBM, Burroughs, and Xerox. He emphasized that their word processor had captured only 3% of sales, whereas IBM already had 75%. He presented the strategic problem of how they were to survive (Shackil, 1981).

Wang announced the need to make a major technological advance in the product of word processors—a new product model. The early word processors were very hard to use, with displays showing only one line; the operator had to calculate to locate any word in the manuscript for editing. In the product strategy meeting, Wang and his engineers conceived of a new model of a dedicated word processor that would be easy to use, have a full-page display and store and retrieve large documents.

With this functionality in mind, the first stages of the project were to define engineering specifications and design the system architecture. Then, programming of the system and application level software began. Wang and his engineers decided to build the word processor as a distributed system around a minicomputer as the system control device. Each application user (word processing person) would have his or her own dedicated terminal for data entry, editing, and recording. Each terminal would be connected to a central minicomputer and printer. Microprocessors would be used in each terminal, in the minicomputer, and in the printer to distribute computation for speed and memory handling.

The project was scheduled to show a product at the International Word Processing Show in New York on June 21, 1976. This gave the team nine months in which to design the machine. As one of the project members, Edward Wild, recalled: "We worked until 10 p.m. almost every weekday, many Saturdays, and some Sundays. I can remember eating lots of McDonald's hamburgers during those long hours" (Shackil, 1981, p. 30).

The hardware parts of the system needed to be designed or purchased. The software needed to be written: system and word processing control. It was the software design that was pushing the state of the engineering art and gave the engineers and computer programmers the most problems:

> Every new design has its share of "war" stories, and even though this system's architecture was relatively simple, the designers' unfamiliarity with microprocessor-based designs did generate some new headaches. . . . There were microprocessor documentation problems. . . . There were elusive system problems. . . . Another problem that arose later on concerned the allocation of space of the rigid disk while a read or write was going on.
> —(Shackil, 1981, p. 31)

Finally, the design was finished, the technical problems solved, and the product got to market on time:

> The opposition was caught flat-footed. Wang leapfrogged IBM in word-processor sales. From being a nobody, the company shot up to capture more than 50% of the dedicated word-processor market.
> —(Shackil, 1981, p. 29)

Introduced in 1977, sales of the word processor began at $12 million, rose to $21 million in 1978, then to $63 million in 1979, $130 million in 1980, and $160 million in 1981. But 1981 was the peak, and also the beginning of the end. The personal computer market had begun and, by 1985, had taken over the word processor application. The Wang company struggled to redefine its position during the 1980s. But, after An Wang's death, the company was bankrupt by 1990.

ENGINEERING LOGIC

The logic of engineering focuses around three kinds of activities: invention, design, and problem solving. Invention is the activity of creating a new technology. Design is the activity of creating the form and function of products, production processes, or services. Technical problem solving is the activity of making technologies work and work well.

Technological Invention

Often, in design or in solving a technical problem, an engineer needs to invent a new way of doing something. Invention is the creation of new technology. The logic of invention is an idea that maps functional logic to physical morphologies. Invention is always new to the person inventing it. Invention need not be new to everyone, as long as the invention solves the immediate technical problem. But if the invention *is* new to everyone, it may be patented.

Design

In design, the essential logic is to create morphological and logical forms to perform function. Form and function are the basic intellectual dichotomy of the concept of design. The logical steps involved require, first, determining the performance required for the function for the customer and then, creating an integrated logical and physical form for fulfilling the function.

In product design, the first logical step is to establish customer needs. This is a list of:

- The customer's applications of the engineered product
- The functional capability of the product for these applications
- The performance requirement for the applications
- The desired features of the applications
- Size, shape, material, and energy requirements for the applications
- Legal, safety, and environmental requirements for the product
- Supplies for and maintenance and repairability of the product
- Target price of the product

Once the customer needs list is established, the next logical step is to establish a "product specification set." These product specs translate customer needs into technical specifications that guide the engineering design of the product.

The third logical step is the design of the product, using ideas from previous product designs along with innovative new design ideas to create a product that meets the product specs and customer needs.

While these logical steps sound sequential, in practice, successful design requires concurrent interactions with marketing and finance and redesign loops as the both the needs and specs get refined into design details. Thus, in a large organization, design occurs in groups of designers and goes from a conceptual design stage into a detail design stage and back and forth until a final design is realized that is ready for testing.

After a design goes into testing, the design must often be modified to correct flaws in the product's design. Once a tested design is ready to be produced, then the design must again be altered to become manufacturable in volume with high quality and to meet target costs. (As much of the manufacturability criteria as possible should be brought early into the design process, to minimize redesign for manufacturability.)

Technical Problem Solving

Engineers make the technologies that a firm uses work, and all technologies are touchy. Nature is always more complicated than the use we make of it. Simplifying nature to make it mostly work for us technologically always stimulates technical problems, which engineering must solve.

The logic of problem solving includes:

- Recognition of a problem
- Identification of the problem
- Analysis of the problem
- Solutions to the problem
- Testing of the solution
- Improvement of the solution and/or redefinition of the problem

In a large organization, problem recognition may not be simple. This requires leadership realizing that there is a problem and acknowledging the existence of the problem. If leaders will not recognize that a problem exists, personnel cannot work on solving the problem. Problems may not be recognized because the leadership does not have the expertise to recognize the problem or because it is politically inconvenient or embarrassing to leadership to acknowledge that a problem exists.

Once a problem is recognized, the next logical step is to identify the nature of the problem, its location, and the client to whom it causes difficulties. Identification of a problem can then logically be followed by analysis of the problem, its sources, and its causes. Once the source and cause of the problem are known, then solutions to the problem can be diagnosed or invented. The proposed best solution can then be tested to see if it solves the problem. If testing shows that the problem is still not solved, then further refinement of the solution or alternative solutions may be tried and tested.

Sometimes, the problem even requires redefinition, if testing shows that the problem was not properly understood initially.

Problem solving is also an essential aspect in the design and invention activities of engineering. There will always be problems in inventions and in new designs that are found only after use in the field. For example, Eric Von Hippel and Marcie J. Tyre examined problem solving in two cases of designing new process equipment (von Hippel and Tyre, 1995). They found that, in about half the problems encountered, information about the potential problem did exist with the users but were not communicated to the designers, as they were not thought to be relevant. In the other half of the problems, the problems appeared only after use of the new equipment in the field.

Since technical problem solving is a major activity of engineering, understanding of the nature of problems and solutions for problems is a critical skill of engineers. This is why science is a base knowledge for engineering, because science provides the knowledge of nature that underlies technology problems.

Case Study: Vannevar Bush, an Engineer of the Twentieth Century

This case study illustrates the dimensions of the career of a professional engineer. Its historical setting is the early part of the twentieth century, when engineering had begun to emerge as a profession. In the United States, Vannevar Bush became one of the most widely known engineers of the twentieth century, playing a key role in mobilizing research for the U.S. World War II effort in the 1940s and in establishing U.S. science and technology policy after that war. At Bush's death in 1974, Jerome Wiesner (who was then science adviser to President Kennedy) stated:

> No man has had greater influence in the growth of science and technology [in the United States] than Vannevar Bush.
>
> —(Zachary, 1995, p. 68)

Bush was born in 1890 in Everett, Massachusetts. He was a descendant of the seafaring traders and whalers of Cape Cod, and valued their earlier culture of independence and entrepreneurship. He attended Tufts College in Massachusetts, graduating in 1914. He then enrolled in Harvard University and, in 1916, earned a joint doctorate in engineering from Harvard and the Massachusetts Institute of Technology (MIT), both in Cambridge, Massachusetts. The First World War was being fought; Bush taught courses in electrical engineering at Tufts and developed a submarine detection device: "To his chagrin, the Navy never even tried it in battle, though it looked promising in tests. (Zachary, 1995, p. 68).

In 1919, Bush joined MIT's electrical engineering department, where he taught and invented for the next two decades. Among his inventions were improved vacuum tubes, a network analyzer, and a mechanically-based differential analyzer (which could solve differential equations in a matter of minutes). He also consulted for industry and helped start several companies, the most successful of which was Raytheon Company.

It was the differential analyzer, a mechanical forerunner of the electronic computer, that brought Bush his initial fame. The differential analyzer used ratios of gears to perform differential calculations in calculus:

> The press gave rave notices to Bush's differential analyzer, which was dubbed a "thinking machine" or "mechanical brain." Scientists were awestruck, too. The president of the National Academy of Sciences called the differential analyzer "the most complicated and powerful mathematical tool ever devised.
>
> —(Zachary, 1995, p. 68)

Bush was awarded the prestigious Levy Medal of the Franklin Institute in 1928 and the Lamme Medal of the American Institute of Electrical Engineers in 1935. In 1939, Bush became president of the Carnegie Institution in Washington, D.C. Eighteen months later, he began advising the U.S. president, Franklin D. Roosevelt, on technical matters for the war preparations at that time.

During the war, Bush ran the Office of Scientific Research and Development, which, in 1944, supported 6000 researchers across the United States at $3 million a week. This was the beginning of a major increase in U.S. funding of research. Bush found that the military then had more respect for scientists than for engineers:

> The [military] brass, he later wrote, considered the engineer to be vaguely disreputable, like some representative of a commercial company on the lookout for business. . . . The "scientist," on the other hand, was regarded as a more disinterested individual and taken more seriously. Bush's R&D office was built on engineers, but given the military's condescending attitudes toward them, he acted as if he hired only scientists.
>
> —(Zachary, 1995, p. 67)

As the war drew to a conclusion, Bush wrote an influential report for U.S. S&T policy: "Science—The Endless Frontier." It endorsed the concept that the federal government's role should be confined to basic research and those technologies essential to national defense. And, with the exception of medical and agricultural research, this was to be the primarily Federal R&D policy for the next 45 years, through the "cold war" with the Soviet Union.

After the war, Bush returned to his post as Carnegie's president. Bush was instrumental in getting the U.S. Congress and President to establish the National Science Foundation, devoted to the support of basic research in U.S. universities. He retired from Carnegie in 1955 and, from 1957 to

1962, served as chairman of the board for Merck & Co., a large U.S. pharmaceutical firm. He continued to invent, designing surgical heart valves, various hydrofoil boats, and other things.

Bush was influential in elevating the concept of an engineer to one of a noble profession. For example, in a speech to the National Academy of Engineering in 1966, he argued that "the professional life of the engineer carries with it so much deep satisfaction, broader even than law, medicine, or science." As G. Pascal Zachary commented:

> Vannevar Bush remains today a towering beacon for his profession. Through words and deeds, he lived out the possibilities unique to electrical engineers—social, political, and economic. . . . His own life demonstrated that an engineer, despite reluctance to engage in partisan politics, could confront the great issues of the times without retreating from the frontiers of knowledge and rationality.
>
> —(Zachary, 1995, p. 65)

ENGINEERING AS A PROFESSION

Engineers regard themselves as belonging to a kind of "profession." A profession is a body of people trained to practice the application of a body of knowledge. Professions organize formal education to master the body of professed knowledge, and they also organize the certification of practitioners in the profession.

In the United States, the organized professions include engineering, law, medicine, nursing, and accounting. Engineers profess to applying bodies of knowledge of scientific, mathematic, and engineering principles to design products or production, solve technical problems, and provide technical services. Engineers have both scholarly societies and a professional engineering society.

In addition to a professed body of knowledge, all professions have codes of ethics, which have to do with practicing the profession responsibly and safely. Engineering ethics stress safety in engineering design. For example, in civil suits involving injury from product use, consulting engineers may testify as expert witnesses about the soundness and safety in a product design.

Thus arises a dual set of loyalties for an engineer—loyalty to professional standards and loyalty to the employing firm. A conflict between these can give rise to the phenomenon called "whistleblowing," when an engineer perceives that management decisions have influenced poor and unsafe design of productions or operations that endanger public safety. The professional expectation of engineering is that the engineer will oppose all unsafe technical practices.

Because of this dual orientation, studies of work performance of engineers have emphasized that management should encourage a "creative tension" for work performance. For example, Denis M. S. Lee suggested that such a creative tension, pushing both technical challenges and business goals, is very important in the development of young engineers, affecting their early and subsequent job performance (Lee, 1992). Also, Gayle S. Baugh and Ralph M. Roberts looked at a sample of engineers as to their perceptions of conflict or complementarity between their professional and organizational commitments (Baugh and Roberts, 1994). They also stress the importance of the employing organization to encourage engineers to see the complementarity of their professional commitments with the organization, and they argued that an important means to do this is for the organization to encourage its engineers to be technically active and participate in their profession.

ENGINEERING DISCIPLINES

Engineers divide into engineering disciplines, centered around generic sets of technologies and their science bases. In the engineering schools of the United States, the traditional engineering disciplines include mechanical, industrial, chemical, electrical and computer and civil engineering. Other disciplines in engineering schools may include specific technologies, such as nuclear engineering.

Engineering disciplines have changed as new generic sets of technologies are invented and innovated into industry. For example, in the United States, the first engineering discipline for which formal university training was provided was civil engineering, begun in the late 1700s at the then new military academy, West Point, for training U.S. Army officers. Mechanical engineering began in the early 1800s, as engineers for the new mechanical industries were needed. After the innovations of the telegraph and electrical power created communications and electrical industries in the late 1800s, U.S. universities began to establish educational programs for training electrical engineers (with the first electrical engineering department in the United States begun at MIT). Later, in the 1920s, after the invention of radio, electrical engineering expanded to include electronic and communication technologies. Even later, in the 1960s, as the computer industry grew, electrical engineering further expanded to add computer engineering. In the early 1900s, as the chemical industry continued to expand from gunpowder to dyes to agricultural fertilizers to plastics, a demand for a new kind of engineer to design and run chemical production led to the training of chemical engineers (with the first chemical engineering department in the United States established at MIT). In the 1980s, chemical engineering departments began to expand to train biochemical engineers as the biotechnology industry grew. With the advance of manufacturing organizations in the twentieth century, the profession of industrial engineering was established to practice the logics of industrial production.

Engineers are trained in the scientific backgrounds for the generic set of technologies they are to practice and in the generic engineering principles of these technologies. These scientific backgrounds are often called the engineering science of a discipline, and the engineering principles consist of sets of generic logics of engineering design and practice. For example, civil engineering builds primarily upon a physical science base of mechanics and hydrodynamics. Mechanical engineering builds primarily upon a physical science base of mechanics and heat. Electrical engineering builds upon a physical science base of electricity and magnetism, and the computer science base of computation. Chemical engineering builds upon the physical science bases of heat and mass transfer, the chemical sciences, and the biological sciences. Industrial engineering has recently built upon the mathematical bases of simulation, scheduling, and optimization.

The engineering disciplines utilized by a firm depend on the kinds of technologies it uses in its core product, production, and service technologies. For example, firms in the chemical industry hire mostly chemical engineers trained in chemistry and chemical engineering. Biotechnology firms need engineers trained not only in chemistry and chemical engineering, but also in biology. Electronic and computer firms hire mostly electrical engineers and computer engineers. Hardware manufacturers hire mostly mechanical engineers, electrical engineers, and manufacturing engineers. Construction firms hire mostly civil engineers.

Supportive technologies used by a firm may be contracted out to consulting engineering firms or to supplier firms. For example, the construction of a chemical production plant will also require mechanical engineering, electrical engineering, and civil engineering. A chemical firm may contract the construction phase to a consulting engineering firm to help design and build the production plant. Afterwards, production will be run and improved by chemical engineers.

ENGINEERING AND CROSS-FUNCTIONAL COOPERATION

Because of an engineer's extensive training in the sciences underlying an engineering discipline, the engineer's world view is, ordinarily, primarily centered on seeing the world as physical nature (or matter). This contrasts with the training of managers in business programs with their intensive training in finance and marketing. The manager's world view is thus primarily centered on seeing the world as capital (or money). Who is correct? Is the world made up primarily of matter or money? Of course, both views are correct, as the world is made both of matter and money. Humans are both physical and social animals. The physical aspect of humanity requires materials for survival, such as food, water, shelter, and clothing. The social aspect of humanity requires cooperation and competition among groups of humans to efficiently produce material requirements from nature. So, the engineer's world view and the manager's world view together could see the world in a balanced perspective—nature and economy.

The important point about the differing backgrounds of engineers and other business function personnel is the special attention that must be paid to ensuring cross-functional team cooperation. In a case study on how engineers interact with marketing personnel, John P. Workman listed typical complaints the two groups make of each other, reflecting their different subcultures. For example, engineers frequently complain that:

- Customers don't know what they want, so what good is marketing analysis?
- Marketing does not have the needed expertise to specify technically sophisticated products
- Marketing's time horizon is too short

In turn, marketing complaints about engineering include:

- Engineers lack perspective.
- Engineers don't appreciate prior customer investments.
- Engineers don't appreciate the diversity of the market segments. (Workman, 1995)

Building cross-functional teams that include engineers and marketing and other business personnel is essential, but difficult. Paul M. Swamidass and M. Dayne Aldridge, in considering the management of cross-functional teams, emphasized the importance of obtaining consensus in decisions and of keeping a cross-functional team focused on deadlines (Swamidass and Aldridge, 1966). They proposed that project leaders should take special care to develop a team perspective and concern for meeting time and budget targets. Thus, in managing the engineering function, special attention must be paid to getting effective cross-functional cooperation operating in effective teams.

ENGINEERING CAREERS

Because of the technical backgrounds involved, engineering careers can follow two paths: a technical path in engineering (or management within the engineering function) or a general management path. In the technical path, an engineer stays within the engineering function of the firm. In the general management path, an engineer moves from the engineering function into other business functions. The career movement of an engineer into general management requires the engineer, by both experience and continuing education, to gain a sophisticated understanding of other business functions in addition to engineering.

ORGANIZATION OF THE ENGINEERING FUNCTION

Engineering is often organized as a separate division of a company, which fosters the specialization of expertise in engineering. However, technical projects that involve engineers should be cross-functional and include other personnel. In organic types of organizations run in a matrix mode, an engineering department may only provide personnel slots, without providing budgets.

SUMMARY

An engineering function is necessary in a firm for the creation and implementation of the technical bases of the enterprise. Since all products have finite model lifetimes, engineering has the responsibility of designing new products and new product models and improving the production of products. Engineering is directly responsible for technological innovation, product and production design, and providing technical services. The logic of engineering focuses on invention, design, and problem-solving in these activities.

Engineering is a profession consisting of educated personnel practicing the application of a body of technical knowledge based on engineering sciences. The disciplines of engineering are organized according to technology and to science base. Engineering careers can follow either a technical path or a general management path. The engineering function can be organized as a separate business division or in a matrix structure.

FOR REFLECTION

Choose an engineering profesion, and describe the change in the profession as science and technology progressed from the origin of the profession.

___16
RESEARCH FUNCTION

CENTRAL CONCEPTS

1. Objectives of corporate research
2. Profit-gap analysis
3. Institutionalization of scientific technology
4. Careers of research personnel
5. Research organization
6. Research and the new product pipeline
7. Funding research
8. Evaluating research

CASE STUDIES

Origin of the General Electric Research Laboratory
William Coolidge, an Industrial Scientist at GE
Xerox Pioneers the Paperless Office
Apple Fails to Dominate the Personal Computer Market
Intel Begins Corporate Research

INTRODUCTION

We will now review the other technical function of a business, the research function. Corporate research and development (R&D) is an asset for long-term competitiveness. Corporate research should be focused on both maintaining existing businesses and preparing the corporation for future businesses. Accordingly, research activities can be classified by their purposes:

1. To support current businesses
2. To provide new business ventures
3. To explore possible new technology bases

The research function is organized in the firm's research laboratories. Technology strategy is budgeted through the research and development budget. The research function is the starting point of innovative new product development. William B. Purdon emphasized the importance of seeing research as a part of the value-adding process of business, with (1) research creating potential value, (2) engineering and manufacturing producing value, and (3) marketing and sales delivering value (Purdon, 1996).

PROFIT-GAP ANALYSIS

Since one of the principal purposes of corporate research is to create and extend the lifetimes of the company's products, anticipating the need for R&D support for products is an important element of research strategy. When product technologies are changing, product lifetimes will be short. In products with mature technologies, product lifetimes are long. But even in a long-lived product, periodic product reformulations do occur to meet changing conditions in market, environment, safety, and the like. Also, to maintain a long-lived product, manufacturing quality must be continually improved and reduced in cost.

A useful way for management to track the attention required for product development is the so-called profit-gap analysis. Figure 16.1 plots the projections of contributions to a business's profits from each of its product lines. Some product lines, such as Product A, can be projected to continue to contribute steadily to profits; others, such as Product C, already have sales dropping and will be obsolete due to aging technology; the company plans to terminate product line C. In between these extremes, some product lines, such as Product B, can be revitalized by redesign.

When the profits from these three types are summed to the total profits of the company, the chart shows the projected profits of the company. If management then plots on this chart a dotted line of its desired profits, the area beneath the dotted line and above the summed profits line shows the gap between desired profits and actual profits. This is the anticipated "profit-gap" unless new products are introduced (Twiss, 1980).

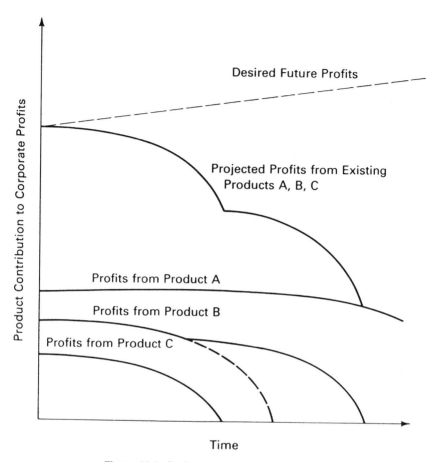

Figure 16.1. Profit-gap analysis of product lines.

The profit gap displays the magnitude over time of the business need for new products.

Case Study: Origin of the General Electric Research Laboratory

This case study illustrates how corporate research revitalized a major product line of a firm, GE, to protect its future. The historical context is the early part of the twentieth century, when the first corporate research laboratories were started. The Research and Development Center of the General Electric Corporation was begun in 1900. Almost a century later, in the 1990s, it still existed, located on a site overlooking the Mohawk River, at the outskirts of Schenectady, New York. In 1982, the GE Center employed about 2100 people, and its budget was about $150 million. GE's total research and development budget was then $1.5 billion, about 48% from company resources and 52% from U.S. government defense contracts. This was still the time of the "cold war" between the U.S. and the Soviet Union. After the cold war ended with the dissolution of the USSR, U.S. firms began to turn away from heavy military R&D contracting and focus more on civilian R&D. The GE Center, however, was originally established for civilian R&D.

The General Electric Company traced its roots to the Edison Electric Light Company, which was established in 1878 to commercialize Thomas Edison's carbon-filament light bulb invention. In 1900, General Electric's principal business was still electrical lighting, producing carbon-filament light bulbs. GE then had an engineering department, and one of its employees was an electrical engineering genius, Charles Steinmetz (Wise, 1980). Steinmetz was a pioneer in developing electrical theory for alternating-current power systems.

Steinmetz was monitoring the progress of contemporary science and technology for GE. At that time, two new kinds of lamps had been invented. One of these was a "glower" lamp, invented by Walter Nernst, a chemist in Germany. This was the prototype of what we now call fluorescent lamps. The second was a mercury vapor lamp invented by an American, Peter Cooper Hewitt. Steinmetz understood that these two lamp technologies were a potential threat to GE's carbon-filament light bulb business.

On September 21, 1900, Steinmetz wrote a letter to Edwin Rice, vice president and head of GE's manufacturing and engineering. Steinmetz told him of the competitive threat to the carbon-filament lamp from both the glower lamp and the mercury vapor lamp. He also pointed out that GE's major competitor already had commercial rights to the mercury vapor lamp. To counter these threats, Steinmetz proposed the establishment of a research laboratory to complement engineering and to search for ways to defend GE's light business. Steinmetz had already discussed this with two other of GE's technical leaders, Albert Davis and Elihu Thomson, who agreed with him. Davis was a patent attorney for GE, and Thomson was a consultant to GE. Thomson had been a founder of one of GE's predecessor companies. He was also a prolific inventor, and a former teacher and long-time mentor to Rice.

Rice took their advice and secured the approval of GE's president, Charles Coffin, to start a research laboratory:

> Rice recognized . . . that the concept of the laboratory had to be broader than mere support for GE's lighting business if he wished to recruit first-rate researchers. Accordingly, as Rice was later to write: "We all agreed it was to be a real scientific laboratory."
>
> —(Wise, 1980, p. 414)

To direct the new laboratory, Rice hired a young assistant professor of chemistry from the Massachusetts Institute of Technology, Willis R.

Whitney, who had been doing research at GE's largest manufacturing works two days a week. Whitney resigned from MIT and began recruiting scientists for the new lab. He began by examining GE's lighting technology. The first task Whitney set for the laboratory was to create a new filament to replace Edison's carbon filament.

One of the researchers Whitney hired was William Coolidge, to whom he assigned the task of examining tungsten as a possible lamp filament. Coolidge succeeded in inventing a new lamp filament from tungsten, and a manufacturing process for it. GE changed its entire lamp production to the new filament. In five years, the new research laboratory saved GE's lighting business. Even today, tungsten remains the material of choice for most incandescent lamps.

Throughout the twentieth century, the GE R&D Center significantly contributed to keeping GE a high-technology company: "Over the years, GE had consistently led all other companies in obtaining U.S. patents, becoming in 1979 the first firm in history to pass the 50,000th patent milestone" (General Electric, 1980, p. 22).

On the importance of scientists in industry, the historian G. Wise has noted:

He [Whitney] was not the first professional scientist to be employed in American industry—or even in General Electric. Nor was his laboratory the first established by that company or its predecessors. That honor must be reserved for Thomas Edison's Menlo Park. But Edison's focus was on invention. Whitney's effort marks a pioneering attempt by American industry to employ scientists in a new role—as "industrial researchers." . . . This laboratory succeeded because it created a new role for professional scientists—a blend of research freedom and practical usefulness not available before 1900. . . . Science and industry were not independent of one another in the United States before 1900. But their contacts then were more limited in scope, intermittent and irregular than they are today.

—(Wise, 1980, p. 410)

INSTITUTIONALIZATION OF SCIENTIFIC TECHNOLOGY

The inclusion of science into the functions of business, in the corporate research laboratory, was a key event in institutionalizing scientific technology in society. The first industrial research laboratories were created in the new technologies of the electrical and chemical industries. In addition to the GE example, many other now-famous corporate laboratories in the U.S. were begun in the early twentieth century, such as the DuPont Laboratories, Bell Labs of AT&T, the Dow laboratory, GM's technical center, and many others. Industrial laboratories were also established in Europe, such as I. G. Farben and Siemens. As the twentieth century proceeded, research laboratories became an important feature of high-technology firms throughout the world. For example, in 1989, the NEC Corporation in Japan spent 10 per cent of its sales on research and development, performing R&D in the corporate research laboratory and in divisional laboratories distributed in the 190 businesses of NEC (Uenohara, 1991).

In 1992, the United States had about 1 million R&D engineers and scientists, with Japan having about 800,000 and Germany about 600,000. The fraction of gross domestic product spent for R&D was around 2.5% for the United States, Japan, and Germany (Geppert, 1994). In the United States, an industrial organization concerned with R&D, the Industrial Research Institute, periodically sponsors surveys of the state of industrial R&D (see, for example, Whiteley et al., 1994).

Historically, in the United States, Herbert Fusfeld identified three critical periods in the evolution of industrial research: 1870–1910, World War II, and the new period of declining technical self-sufficiency starting in the 1970s.

In the transition period from 1870 to 1910, new industrial research organizations began to give the new science-based chemical and electrical industries timely technical support:

> The companies no longer had to depend on the unpredictable advances generated externally. From this start . . . there was a steady growth of industrial research up to World War II.
>
> —(Fusfeld, 1995, p. 52)

Next, the experiences of military research in the United States during World War II had consequences for subsequent industrial research:

- A great reservoir of technical advances became available for further development in commercial areas.
- Public expectations were raised for the potential of science and technology in new products.
- New techniques of systems development were successful in planning and conducting complex technical programs that required the generation of new knowledge as an integral part of the planned program.

> —(Fusfeld, 1995, p. 52)

Immediately after the war, U.S. research and industry dominated the world until the war-devastated economies in Europe and Japan were reconstructed. By the 1970s, management in the United States felt the consequences of the industrial progress by the rest of the world. Then they realized that technical self-sufficiency within a U.S. company was no longer practical.

> For the 30 years following World War II, [U.S.] corporations were able to plan growth strategies based on technical resources that existed internally . . . That situation changed. . . .
>
> —(Fusfeld, 1995, p. 52)

After the 1970s, the conditions of the growing global economy became widely recognized. Corporate research in a global economy required new characteristics, such as:

- Strategic use of external resources from outside the company
- Dispersion of corporate technical activity beyond a central laboratory
- Emphasis on the effective integration of research into total corporate technical resources
- Major effort on the organized pursuit of technical intelligence world-wide
- Extensive foreign-based R&D activity to support growth in international markets
- Improved integration of technical strategy into corporate business strategy (Fusfeld, 1995)

In summary, the establishment of corporate research laboratories helped institutionalize scientific technology in the twentieth century. Over that time, the nature of the corporate research laboratory continued to change as management struggled with how to optimize research's contribution to corporate profitability and long-term survival. For example, even in the GE corporate laboratory, management was still changing its perspectives in 1995. Lewis Edelheit, then GE's senior vice president for corporate R&D, commented:

> How can a corporate-level R&D lab renew its role as a vital part of a winning business team? My answer is . . . cost, performance, speed, quality. In the past, innovation often began with a performance breakthrough made at the corporate lab. . . . The result was relatively slow development of often costly products that nevertheless might prove winners in the marketplace because they offered capabilities no competitor could match. That does not work today . . . because technology is exploding worldwide. Dozens of labs are at the forefront today whereas yesterday there might have been only one or two.
> —(Edelheit, 1995, p. 14)

CAREERS OF RESEARCH PERSONNEL

Both engineers and scientists are hired to perform the research function. Scientists differ from engineers by their training and professional orientation. The transformation from a scientist to an industrial scientist is even more drastic and difficult than that of the engineer. Engineering education aims to train engineers for industrial employment; scientific education usually aims to train scientists for academic employment. Therefore, the transition of a scientist into an industrial setting is the more difficult of the two.

What scientists and engineers share in their training is the view that the world is made up of physical matter in nature. However, scientists are trained only to discover and understand nature and not to manipulate it, as do engineers. Thus, to become industrially oriented, scientists must learn both the engineer's perspective on the value of manipulation and the manager's perspective on the economic value-addedness of manipulation.

Due to the business context of the firm, the career of an industrial scientist must involve a creative tension between research and management. Industrial research personnel must be specially recruited, motivated, trained, and integrated into the corporation. The balancing of business with research requires focusing fundamental knowledge on invention and technical problem solving, which increases the productivity of the firm. Accordingly, this dual aspect of research personnel requires a dual basis for performance evaluation. Donald K. Wilson, Roland Mueser and Joseph A. Raelin have noted the need for evaluating researchers on two sets of criteria, one technical and the other managerial (Wilson et al., 1994).

To accommodate this duality to individual scientific careers in industry, some corporations have instituted what is called a "dual career ladder." In a dual career ladder, there is a technical route for advancement for research engineers and scientists to remain researchers for the duration of their employment, or alternatively, a managerial route for research scientists and engineers to become general managers. Not all researchers should or can become managers. A dual career ladder, with both technical and management sides, allows some technical people to advance without becoming general managers. However, the effectiveness of the dual ladder approach remains a subject of debate. For example, Thomas J. Allen and Ralph Katz pointed out that it is the PhD group of scientists and engineers who appreciate having the choice of a dual career ladder; those who are engaged in fundamental research most likely choose to remain on the technical career ladder

—(Allen and Katz, 1992).

Case Study: William Coolidge, an Industrial Scientist at GE

This case study illustrates the career of a successful industrial scientist, William Coolidge. We recall that it was Coolidge who, as a young scientist, was critical to saving GE's lamp business.

Coolidge was born in 1874, the son of a farmer and a shoe factory worker. He attended a rural one-room elementary school. Graduating from a small high school as an outstanding student, he went off to college in 1891, which then was still unusual. Moreover, he enrolled in a new electrical engineering program at the Massachusetts Institute of Technology (Wolff, 1984).

Coolidge graduated with a bachelor's degree and decided to go on and become a scientist. But in the 1890s, U.S. universities still were not training scientists. To get a real scientific education, a student had to go to Europe. Coolidge went to Germany and returned in 1905 with a doctorate from the University of Leipzig. He then had a $4000 debt for his graduate education. He took a position at MIT as an assistant to Arthur Noyes, a notable American scientist. Coolidge's pay as an assistant was $1500 a year.

Willis R. Whitney, who was then the new director of GE's research laboratory, had taught the young Coolidge in a chemistry class when Whitney was a chemistry professor at MIT. Hearing that Coolidge had returned with a German doctorate, Whitney offered Coolidge a job in GE's new laboratory for $3000, double Coolidge's MIT salary. Since this would allow Coolidge to pay off his education debt more quickly, Coolidge was attracted by the offer. Yet he worried whether he could conduct scientific research in industry. Whitney assured him he could, as this was the mission of the new GE laboratory. The only constraint was that the scientific research needed to be related to GE's commercial interests.

Coolidge accepted the offer and moved from Boston to Schenectady, New York, to work at GE's facility. There, he was pleased at the atmosphere for science that Whitney was creating: "Dr. Whitney had already successfully transplanted a lot of academic atmosphere from MIT to the new laboratory. . . . The misgivings I had about the transfer all proved to be unfounded" (Wolff, 1984, p. 81).

We recall that Whitney had a research strategy to find a replacement for the carbon filament. He assigned to each of his researchers the responsibility for exploring each of the elements in the periodic table that was a metal with a melting point equal to or higher than that of tantalum. One of these was tungsten, which Whitney assigned to Coolidge. Coolidge wrote to his parents: "I am fortunate now in being on the most important problem the lab has ever had. . . . If we can get the metal tungsten in such shape that it can be drawn into wire, it means millions of dollars to the company" (Wolff, 1984, p. 84).

The technical problem of using tungsten as a filament was its natural ductility. It had a high melting point but was brittle—too brittle to draw into the thin wire shape needed to perform as a lamp filament. In June 1906, Coolidge had his first research break, observing that mercury was absorbed into hot tungsten and, on cooling, formed an amalgam. Next, Coolidge experimented in making other amalgams of tungsten. He observed that he could make one with cadmium and bismuth absorbed into tungsten. This amalgam could be squeezed through a die to make a wire. Coolidge was on the right path. In March 1907, Coolidge discovered that, by heating this amalgam to 400° Fahrenheit, it could be bent without cracking.

Coolidge had a potential candidate material for a new filament; the next step was to figure out how to manufacture it. He decided to visit wire- and needle-producing factories and toured several in the New England region. There he saw the swaging techniques used for wire making.

Swaging was a process of gradually reducing the thickness of a metal rod by repeated use of dies to draw smaller and smaller diameters. In May 1909, back in Schenectady, Coolidge purchased a commercial swaging machine and altered it to swage his tungsten amalgam into wire. It did make tungsten wire, but when the wire was heated in a lamp as a filament, it broke. Coolidge examined the failure and learned that it was due to crystallization of the tungsten wire upon heating.

This problem was solvable. Coolidge knew that, in a similar problem in making ice cream, glycerine was added to prevent the forming of ice crystals as the milk froze. Using this analogy, Coolidge tried adding another substance to the tungsten amalgam, thorium oxide, to prevent crystallization. It worked. On September 12, 1910, Coolidge achieved a technical success. He had developed a new tungsten wire for a lighting application and a manufacturing process to produce it.

GE's management was pleased, and immediately innovated the new material and process. By 1911, GE had thrown out all the earlier lamp-making equipment and was producing and selling lamp bulbs with the new tungsten filaments. The research effort had cost GE five years and $100,000. But by 1920, two-thirds of GE's $22 million profit came from the new lamps.

After Coolidge's innovation of tungsten filaments in incandescent lights, he turned his scientific attention to a related but entirely new area: production of X rays. Coolidge had the idea that tungsten might be a useful material as a target in an X-ray tube. With the help of another GE scientist, Irving Langmuir (who would later receive a Nobel prize for his discoveries in physics), Coolidge invented the first practical X-ray tube in 1913. Over the course of the next 15 years, Coolidge continued to make technical contributions to X-ray applications. In 1928, he was promoted to associate director of the GE laboratory, beginning a research management career.

The 1930s began with an economic depression in the United States, and GE was severely stressed. By April 1932, Whitney had decided to retire as director of the laboratory. His health had begun to suffer from worry about keeping the GE lab going and fears that it might be shut down in view of the bad economic conditions. On November 1, 1932, Coolidge succeeded Whitney to become the second director of the GE laboratory. Immediately, Coolidge took action to reduce expenses and save the lab. Coolidge put the laboratory on a four-day work week and cut the work force from 555 to 270 people. He was careful to prune, keeping the best researchers and maintaining the emphasis of the laboratory on fundamental research and the invention of new technologies.

It was tough to maintain morale and to continue the search for new products. By December 1933, Coolidge had convinced higher management to permit him to add five new chemists to develop chemical-based products. Research continued through the great depression in the United States, and many of the inventions of the 1930s provided GE with a base for its war effort in the 1940s and economic expansion thereafter.

In terms of management style, his contemporaries saw Coolidge as combining expense control with a leadership of technical excellence. He interacted with his staff and was accessible. He walked around the lab, looking, as some said, "more like a scientist than a laboratory manager" (Wolff, 1984). He was encouraging to his researchers, with comments such as: "This is fascinating; anything we can do to help?" One of his colleagues said of Coolidge:

> That man oozes optimism of an inspiring brand. You feel in his presence that if all things are not possible, many are. Yet he has plenty of circumspection, with a verifying, sagacious mind that readily isolates what is either impossible or extraneous.
>
> —(Wolff, 1984, p. 84)

RESEARCH ORGANIZATION

There are three ways to organize research:

1. Divisional laboratories reporting to business units
2. A corporate-level laboratory
3. Both divisional laboratories and a corporate-level laboratory

Ordinarily, research in divisional laboratories is focused on next-product model design and on production improvement, whereas research in corporate laboratories is focused on next-generation product lines and on developing new businesses from new technology.

The kind and number of research units depend on the size and diversity of businesses in a firm. A single-business small firm will likely have only an engineering department. A medium-sized firm will likely have an engineering department and divisional laboratories. A large, diversified firm will likely have engineering departments and divisional laboratories in business units, and also a corporate laboratory for all businesses.

Research organization varies by industry. For example, the first annual Industrial Research Institute R&D survey in the U.S. noted that "the art of technology management may be more industry-specific than some observers have realized" (IRI, 1994). Even within an industry, research organization varies. Charles Bosomworth and Buron Sage surveyed 26 U.S. firms in 1993 as to their management practices. They found that R&D investments ranged from 0.6% to 15 % of sales and that research organization varied from only a central research laboratory to only divisional laboratories. Also, research strategies varied from emphasizing defensive to offensive technology strategy (Bosomworth and Sage, 1995).

Arthur N. Chester studied the research management practices in aerospace and electronics systems companies (Chester, 1994). He found that most of the 30 largest firms had either one or two corporate laboratories, but there was substantial variation. TRW and General Dynamics (GD) had no central research laboratory, while Hitachi had nine central research laboratories. Firms without central laboratories, such as TRW and GD, depended more on outside technology than those with central labs. Chester also found that the corporate research organization reflected the core technologies and science bases of the relevant businesses. For example, Siemens had 5 technology groups, with 24 core technologies.

Is there a "best" organization for research? Albert Rubenstein addressed this question after having observed the organization of research in many firms over a long period, and his conclusion was that no single organizational form could provide a "best" solution (Rubenstein, 1989). However, the influence of research organization on research activity is clear. Research organization consisting only of decentralized divisional labs encourages a short-term focus, mainly on the current businesses of the corporation. Therefore, in this form, management must provide special attention to focus on long-term issues. In contrast, research organization consisting only of a corporate research lab encourages long-term focus, but at the cost of short-term relevance. Accordingly, in this form, management must pay special attention to making its research relevant to the current businesses of the firm.

Thus, a research organization consisting of both divisional labs and corporate labs has the potential strengths of focusing on both current businesses and future businesses. The problem, however, is that research subcultures develop differently in the divisional labs and the corporate labs. This difference can foster competition, rather than collaboration:

> As the operating divisions begin to flex their decentralized muscles and start acting as though they were indeed independent enterprises, . . . they begin to become impatient with the level or quality or relevance of the work being done in the central R&D activity. For those division managers who see a real need for strong, direct technical inputs to their division's operation, the central lab seems unwieldy, distant, and not very responsive to their immediate and near-term future needs.
>
> —(Rubenstein, 1989, p. 41)

For example, research at DuPont from 1950 to 1990 was organized as the combination of a central research lab (Central Research and Development), an engineering R&D (part of the engineering department of DuPont), and business-specific R&D conducted in engineering departments in the strate-

gic business units. The central research lab then became increasingly discon-
nected from the business units:

> Where once breakthrough technology was the hallmark of DuPont's research,
> the focus over the four decades increasingly shifted to incremental improve-
> ment of product lines and manufacturing processes. . . . Central R&D became
> more isolated from business management and their work more speculative
> and academic.
>
> —(Titus, 1994, p. 351)

In 1992, research at DuPont was reorganized:

> In effect, R&D is almost as centralized now as it was forty years ago, but for
> different reasons. . . . The old paradigms of sequential development are no
> longer operative in today's competitive global environment.
>
> —(Titus, 1994, p. 351)

In conclusion, a mixed research organization form may not be "best" for
everyone's purposes, but it may be necessary for all purposes. This organizational conflict in centralized versus decentralized industrial research will always exist and cannot be eliminated by any management procedure. Accordingly, this should be constructively managed. Effective collaboration
between research units and between research and business units is very important, as well as are is concurrent practices in R&D.

Case Study: Xerox Pioneers the Paperless Office

This case study illustrates the importance of the proper management of
technological innovation to capture the benefits of corporate research. The
historical setting is the 1980s, in the first decade of new personal computer market.

We recall that several small companies had pioneered the new personal
computers, but the market exploded when the mainframe computer giant,
International Business Machines, entered the new market. The reputation
and brand name of IBM accelerated market acceptance of the personal
computer in the business community. We also recall that IBM neither
did pioneering research in personal computers nor captured the personal
computer market. Although IBM had been a research pioneer in mainframe computers, IBM did not want to pioneer research in personal computers, for IBM's management then cared only about mainframe computers. It took another company, Xerox, to pioneer the research for personal
computers.

The Xerox company had evolved from Haloid in the 1950s, when Wilson developed Carlson's invention for xerography. Wilson guided Xerox
into a giant company, dominating the new copying machine industry
through strong patent protection. But, by 1970, Xerox was a company

with an aging patent-protected technology and many competitors eager to grab a larger share of the market. Xerox's leadership decided it needed a new technology vision and thought of combining copying and the computer into a new concept of a "paperless office." It turned out to be a brilliant long-range vision.

Xerox had earlier bought a computer company, which was chewed up in the mainframe business by IBM in the late 1960s. In the early 1970s, Xerox sold its computer business. But Xerox still had the technological vision of computers, and it set up a new research laboratory in Palo Alto, California, to explore new computer directions for the paperless office concept. (Xerox retained the older research laboratory in Rochester to continue to concentrate on copying technologies.) (Uttal, 1981).

Xerox hired George Pake, a physicist, to head the new laboratory, and he located it next to Stanford University, which was strong in both electrical engineering and computer science. Pake hired many bright young researchers, including an important one named Alan Kay. In the late 1960s, Kay had been influenced by the ideas of an MIT professor, J. Licklider, who envisioned an easy-to-use computer, portable and about the size of a book (Bartimo, 1984). At Palo Alto, Kay's ideas, along with those of his colleagues, blossomed into a new vision of the future computer—personal computers linked in a communications network with laser printers, and operated with icons, desktop metaphors, and mouses. They called this the Altos system, an experimental testbed built at PARC in 1979 for the use of PARC researchers. What did Altos look like in 1980? It looked like what one could eventually see only after 1990 in offices, with local area networks of Apple's Macintosh computers, hooked together with ethernet coupling, using icons, desktops, mouses, and object-oriented programming software. But all these state-of-the-art ideas of distributed PC systems of the 1990s had been invented back then in the 1970s by Xerox's PARC scientists.

Did Xerox make a bundle from its visionary investment in computer technology? No, Xerox did not make a dime from it. As the business reporter, Bro Uttal, commented in 1983:

> On a golden hillside in the sight of Stanford University nestles Xerox's Palo Alto Research—and an embarrassment. For the $150 million it has lavished on PARC in 14 years, Xerox has reaped far less than it expected. Yet upstart companies have turned the ideas born there into a crop of promising products. Confides George Pake, Xerox's scholarly research vice president: "My friends tease me by calling PARC a national resource."
> —(Uttal, 1983, p. 97)

Why this embarrassment by Xerox leadership? Because other leadership in Xerox could not take advantage of this vision. In 1980, the manager of Xerox's office products division had visionary myopia. He saw the future as a network of word-processing stations, which a company called Wang had then innovated. Xerox's office products division put out a lookalike product using some of PARC's innovations. It was called the "Star" workstation, and was a technical and commercial failure. It cost Xerox its entire investment in personal computers, and it failed to capture a market position in the emerging personal computer market. The failure of Xerox's office products division manager was a failure of technological vision to go with PARC's brilliant vision and research.

RESEARCH AND BUSINESS SUBCULTURES

Innovative product design requires a visionary and collaborative partnership between research and marketing. As we saw in the Xerox PARC case, it is possible to create pioneering research without successfully commercializing it. We also saw previously, in the case of IBM and personal computers, that it is also possible for a high-tech firm to neither perform business-relevant creative research nor to implement it.

There are great differences in culture between research labs and business units. Paul D. Klimstra and Ann T. Raphael observed some of the basic organizational differences:

> There are long-standing and deeply-rooted problems involved in integrating R&D and business strategies. . . . R&D and business organizations have conflicting goals and practices.
> —(Klimstra and Raphael, 1992)

Among the conflicting goals and practices, they especially noted the differences with respect to time horizon, finance, product, and methods:

Characteristics	R&D Org	Business-Unit Org
Time horizon	Long-term	Short-term
Finance	Expense center	Profit center
Product	Information	Goods/services
Method	Technology push	Market pull

R&D is generally a long-term investment. Even the shorter product developments take two to three years from applied research to development. The longer developments from basic research usually take ten years. Thus, R&D is fundamentally "strategic" in its planning horizon. In contrast, business units are always under the quarterly profit accounting system, focused principally on the current year's business. Business units are fundamentally "operational" in their planning horizon. Yet, if an operational component is not included in R&D strategy, it will not integrate properly into the business unit's planning horizons. Conversely, if business units do not have strategic long-term planning horizons, they have strategic difficulty in integrating the R&D units' plans.

Another differing characteristic is that, organizationally, the R&D laboratories are expense centers, since they do not directly produce income, whereas the business units directly produce income. This can lead to arrogance by the business units toward the R&D organizations, making it more difficult for R&D to elicit support and cooperation from the business units.

The characteristics of products from the R&D center and business units also create important cultural differences in the organizations. Business units' products are goods and/or services sold to customers of the firms, whereas R&D center products are information, understanding, and ideas that must be communicated internally to business units and then embodied into goods and services. This can generate cultural differences on how the different organizations value "ideas." The R&D unit must place a higher value than the business unit on the "idea" of a future product, whereas the reality of the business unit is focused on the existing products, not as ideas, but as real objects in the marketplace.

Finally, the methods of the organizational units differ, with the R&D unit valuing scientific and engineering methodology and principles that result in "pushing technology" from the opportunities of technical advancement. In contrast, the business unit, with its direct contacts with the market, will be primarily interested in "market demand," meeting the needs of the existing markets.

Thus, the problem of making research relevant to business requires formal procedures to foster cooperation. However, simple bridging mechanisms may not be sufficient. A culture of trust and a history of relevance and creativity must also be built by experience. From the perspective of the researcher, cooperation by a business unit will always be problematic. Robert A. Frosch summarized the researcher's perspective on the customers for the research:

> After 40-odd years of working in application-and-mission-oriented research, I have come to believe that the customer for technology is always wrong. . . . I have seldom, if ever, met a customer for an application who correctly stated the problem that was to be solved. The normal statement of the problem is either too shallow and short-term. . . . What really happens in successful problem-solving . . . is the redefinition of the problem. . . .
>
> —(Frosch, 1996, p. 22)

FORMALIZING RESEARCH AND BUSINESS INTEGRATION

To tackle the profound organizational differences in cultures, it is important to formalize procedures for strategically integrating R&D into business units' activities. For this, Paul D. Klimstra and Ann T. Raphael argued the

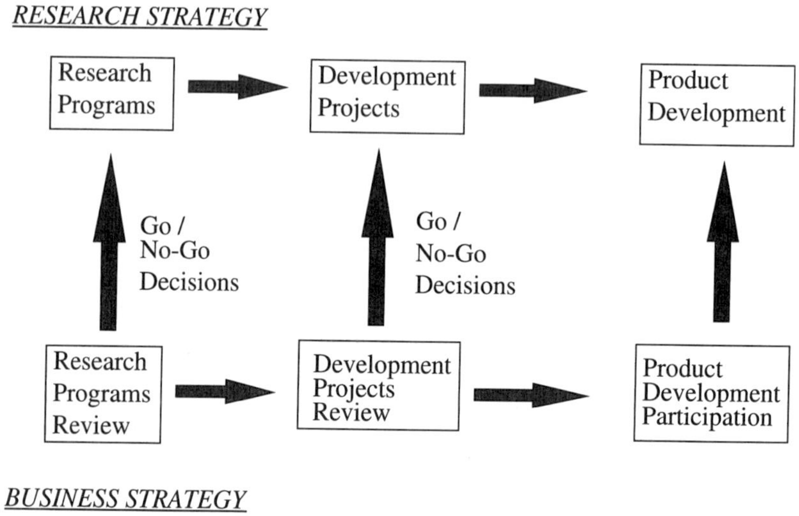

RESEARCH STRATEGY

BUSINESS STRATEGY

Figure 16.2. R&D product pipeline.

usefulness of having two parallel sets of formal decision processes, in what they called an "R&D product pipeline." As illustrated in Figure 16.2, a procedure for an R&D product pipeline involves two parallel sets of activities to integrate research strategy in research labs with business strategy in business units. This requires that, as research programs move into development projects and into product development, research strategy should be formally reviewed by business units as part of their business strategy.

Business units should participate in the research program reviews. Project selection decisions should be made jointly made by research and business units. While development projects are proceeding, business unit participation with the research unit in development project reviews continues to be necessary. Joint go or no-go decisions can be made by research and business units before product development begins. During product development, business unit personnel should participate actively with research unit personnel in joint product development teams.

Case Study: Apple Fails to Dominate the Personal Computer Market

This case study illustrates the need for a research function. It shows the long-term consequences of the failure of top leadership to create an organizational capability to do advanced research for the company's future. The historical setting is again the early 1980s, as the personal computer market began to grow. We recall that Apple was one of several high-tech new ventures launched in the late 1970s as one of the pioneers in the personal computer industry. Apple was the only one of the first eight personal computer companies to survive into the 1990s. But even then, Apple never succeeded in dominating the PC market:

> Apple Computer, once the hip flag bearer of high tech, is in sad decline. There are lessons aplenty.
> —(Rebello et al., 1996, p. 34)

Steve Jobs and Steve Wozniak founded Apple in 1977 with the capital and management assistance of A. C. "Mike" Markkula. Markkula recruited Mike Scott as president of Apple, since neither Jobs or Wozniak had management experience. At first, Apple did well, with an open architecture policy that allowed other companies to write software or market peripheral equipment for the Apple, which gained 27% of the new market by 1981. But in 1982, IBM's entry grabbed 27% of the PC market, matching Apple's share. Moreover, IBM had entered with a technically superior product using a 16-bit microprocessor.

But meanwhile, since the IBM PC was priced higher than the Apple, Apples continued to sell. This gave the company time to respond to IBM's technical challenge. Wozniak took up the task of upgrading the Apple II to keep it marketable, while Scott began the development of a business model, the Apple III, with a 16-bit microprocessor to compete with IBM's PC. But Scott's product development leadership was poor. The Apple III was slow to market and full of bugs, without any superiority to the IBM PC.

Next, Steve Jobs asserted leadership, knowing that Apple needed new technology. But since Apple did not do research, Jobs had to look outside. Jobs heard of Xerox's research in its PARC laboratory and visited it. There, Jobs saw the future (even though Xerox's office product manager could not see it). Recalls Lawrence Tesler, who helped develop Smalltalk: "Their [Apple's] eyes bugged out" (Uttal, 1983, p. 98).

Seven months later, Jobs hired Tesler and launched a new product development, which was eventually to become Apple's Macintosh:

> The Apple II office computer was a bug-infested flop. And Lisa, precursor to the Mac, was an expensive dud. Jobs' masterpiece, however, the 1984 Mac, was a stunner. . . . It was also severely underpowered and limited in expendability. The market balked, and in May 1985, Jobs was pushed out of daily operations.
> —(Rebello, et al., 1996, p. 36)

At first, Macintosh sales were slow, but a new software vendor introduced a new functionality to use the Mac's features and launched desktop publishing. This saved the Mac by bringing in corporate customers. With the desktop publishing application, the Mac succeeded. By 1987, Apple gained a significant place in corporate sales and expanded around the globe. In 1987, the MS-PC world was technologically far behind the Mac. Yet, because of the marketing clout of IBM and its clones, the MS-DOS world continued to dominate PC market share. And this would continue because of the loss of previous software investment if users changed operating systems.

Jobs became chairman of the board and searched for a president to succeed him. He picked John Sculley, who had 20 years of marketing experience at Pepsico:

> But Apple [under Sculley] entered the 1990s with an overpriced product line and a bloated, over-perked executive staff. Microsoft Windows was gaining ground and Apple's rate of innovation was slowing. . . . Determined to catch the next technology wave, Sculley put himself in charge of research and development—and came up with the Newton personal digital assistant, a marketing and technical fiasco.
>
> —(Rebello et al., 1996, p. 36)

In June 1993, the board of Apple fired Sculley. Markkula next chose Michael Spindler, head of the successful European division of Apple, to become CEO: "Michael Spindler started off with a 2,500 employee layoff, the first move toward a new, low-margin business model." (Rebello et al., 1996, p. 37). He produced inexpensive Macs for the home market and introduced a new higher-power-mac line with a new chip, the PowerPC chip, developed jointly with IBM. But, in 1995, Apple stumbled dramatically, when Spindler's large inventory of lower priced and lower-powered Macs were ignored for the PowerPC line and there were not enough produced for the Christmas season sales:

> The [1995] Christmas quarter was a disaster . . . January 1996 brought news of a last quarter loss of $69 million. Laying off 1,300 workers is just the first step in an overhaul that could include Spindler's ouster and/or even a sale of Apple.
>
> —(Rebello et al., 1996, p. 37)

Markkula called an emergency meeting for January 31, 1996, at the St. Regis Hotel in New York City. Spindler was surprised at the request for his resignation and argued for more time:

> The board was firm: Spindler had contributed much over his 16 years at Apple, but directors had been surprised by plunging gross margins, throwing into question management's credibility.
>
> —(Armstrong and Elstrom, 1996, p. 29)

The Apple board selected Gil Amelio of National Semiconductor to replace Spindler. Amelio had transformed National Semiconductor from its worst loss in 1991, $151 million, to a best-year profit in 1995 of $262 million. Amelio finally began shopping for a new operating software system:

> Apple was known to be casting about for a new operating program, the software that serves as the computer's master-control panel. Apple's in-house development effort, code-named Copland, had collapsed. For Apple, shopping for an operating system was a humiliation akin to General Motors' having to buy engines from another company.
>
> —(Lohr, 1997, p. 16)

Earlier, when Steve Jobs had been ousted from Apple, he had set up a company, Next, to develop a next generation PC. The hardware Next produced had not been successful, but the operating system was advanced:

> In recent years, Next had become a company with excellent technology but shrunken ambitions. . . . The company was scaled back to provide software for computer programmers, especially those building Internet sites.
>
> —(Lohr, 1997, p. 16)

When Jobs heard that Apple was shopping for a new operating system, he proposed Next's system:

> [On Dec. 2 in Apple's corporate headquarters, in Cupertino, California] Jobs and the three executives from Apple gathered in the eighth-floor conference room next to Amelio's office. . . . Jobs [made] his pitch, explaining why Next's operating system was Apple's best choice. The Apple executives were impressed.
>
> —(Lohr, 1997, p. 16)

Apple had turned back to the only executive it ever had who had vision to keep bringing new technology into Apple:

> On Dec. 20, 1996, Apple's C.E.O. and chairman, Gilbert F. Amelio, announced that the company would buy Next Software Inc. for $400 million. For that price, Apple [gets an advanced operating system software and] also gets Steven P. Jobs. . . . So Jobs becomes the computer era's prodigal son: his return to Apple after more than a decade in exile is an extraordinary act of corporate reconciliation, a move laden with triumph, vindication and opportunity.
>
> —(Lohr, 1997, p. 15)

This case illustrates the results of the failure to create a capable corporate research laboratory. Although Apple aspired to be a technology leader, successive leaders did not establish a scientific technology capability in the firm. Apple coasted too long on borrowed, once-innovative technology (which Jobs had found in 1980 in Xerox's PARC laboratory). By the mid-

1990s, Apple had not become a technology creator, only a technology borrower, which was great in the short run but not in the long run.

On July 10, 1997, after only 18 months on the job as Apple's CEO, Gilbert Amelio resigned after a "confrontation with the company's Board of Directors over the company's faltering performance" (Corcoran, 1997, p. E1).

Next on August 7, 1997, Apple announced that Microsoft was investing $150 million in Apple for 7 percent ownership:

> Apple Computer Inc. yesterday accepted what amounts to a $150 million aid package from its historic rival, software giant Microsoft Corp. . . . The two companies unveiled an alliance in which Microsoft agreed to . . . jointly develop new technologies . . . and expand the lineup of Microsoft software for Apple's Macintosh computers.
> —(Chandrasekaran and Shannon, 1997, p. A1)

In addition, the board of Apple named Steve Jobs to the board, from which he had left 11 years ago after losing control of Apple to John Sculley. Now back in control at Apple, Jobs arranged the agreement with Gates at Microsoft, which was announced in August at the Apple trade show in Boston. There, some loyal Macintosh users in the audience stated their fears this agreement would mean the end of Apple by being swallowed up by Microsoft. To them, Jobs argued that "the era of setting this (PC operating systems) up as a competition between Apple and Microsoft is over as far as I'm concerned." (Chandrasekaran and Shannon, 1997, p. A1)

And it certainly was over. Microsoft had won the competition. Apple, after all, had allowed Microsoft over ten years, time enough to catch up on the once brilliant Xerox research that Jobs had innovated in the Macintosh computer. Give a rival long enough time to catch up with your technology, and they will. This is why new technology is always only a *temporary* advantage. And this is why a technology leader must *perform fundamental research* to stay ahead of the competitors who are chasing.

In June 1997, a popular business journal, *Business Week,* reported on the top twenty research laboratories in the United States, and Apple was not on it. Yet IBM Research, Xerox PARC, Microsoft Research were on the list (Gross, 1997).

FUNDING CORPORATE R&D

Budgeting corporate R&D is difficult because it is a risky investment over varying return periods. R&D costs are usually deducted as current operating expense and treated as part of the administrative overhead. In practice, most R&D budgeting is done by incremental budgeting, increasing research when business times are good, cutting research when profits drop.

Accordingly, the level of R&D expenditures tends to be an historically evolved number that has depended on many variables, such as the rate of change of technologies on which the corporate businesses depend, the size of the corporation, levels of effort in R&D by competitors, and so on. In areas of rapidly changing technology (high-technology), firms tend to spend more on R&D as a percentage of sales than do firms in mature core technologies: High-tech firms spend in the range of 6% to 15% of sales, whereas mature technology firms may spend 1% of sales or less.

The share of R&D divided between corporate research laboratories and divisional laboratories also differs among industries but, generally, divisional laboratories have the greater share because of the direct and short-term nature of their projects contributing to profitability. For example, in high tech firms, the corporate research laboratory might get as much as 10% of R&D, but seldom more.

In the corporate research lab, R&D funding is usually of three types: (1) allocation from corporate headquarters, (2) internal contracts from budgets of business units, and (3) external contracts from government agencies. The corporate allocation provides internal flexibility for the corporate lab to explore long-term opportunities. The internal contracts provide direct service to business units. External contracts from government agencies provide either a direct business service to government or additional flexibility for the research laboratory to explore long-term future technologies. Normally, internal contracts will provide the majority of corporate laboratory R&D (unless the corporate laboratory is in the government-contract R&D business).

Many studies have tried to find correlations between R&D investments and profitability, without much success. The reason is that not all R&D is good research, and not even all good research will be useful to a firm. The quality of managing the R&D function is more important than the quantity of resources spent on R&D. Thus, some firms that have managed R&D well have received great benefit, while other firms, even with great expenditure, have not benefited from it.

Case Study: Intel Begins Corporate Research

This case study illustrates how a CEO's vision can create corporate research. It also provides an interesting comparison about the quality of CEO leadership with the previous case study about Apple; for while Apple blew its technological lead during the 1980s, Intel strengthened its technological position through several generations of PC chips that increased performance while maintaining the customers' software continuity. The historical setting was in the 1990s, the second decade of the personal computer industry. We recall that Intel became the most successful chip manufacturer in the United States after IBM selected its chip for its first personal computer. IBM had inadvertently established Intel as the industrial standard for personal computer microprocessors.

Under aggressive leadership, Intel continued to design next generations of microprocessor chips as the pace of technological innovation in chip density provided it with new capability. The original chip in the IBM PC was a 16-bit chip, the Intel 8084, which was followed by the 80286, the 80386, the 80486. Finally, Intel introduced a full 32-bit chip in the Pentium model. The Intel and Microsoft chips and operating systems became the standard of the PC industry for 90% of the market. In 1986, Intel's stock sold at $6 dollars a share; in 1992, $25 a share; in 1995, $60 a share; and, in 1997, $140 a share. Intel's profits soared from $0.2 billion in 1987 to $5.2 billion in 1996 (Kirkpatrick, 1997).

Andy Grove became CEO of Intel in 1986, guiding the company in its growth, resulting in Intel's becoming one of the top-earning U.S. companies in 1996. Yet, in the early years of Intel's ride, the company was as an industry technology follower, not a leader. Intel's success was based on technology progress in the chip industry and customer growth of the market. Could this strategy continue to be successful in the 1980s?

> Around 1990, Grove started moving the company from being an industry follower to a leader. . . . The speed of the microprocessor was starting to outpace the performance of the rest of the machine. . . . An Intel division had a proposal for a [new] bus called PCI. . . . Today PCI is the standard bus on PCs.
>
> —(Kirkpatrick, 1997, p. 64)

Intel's success in innovating other parts of the PC stimulated Grove into rethinking Intel's technology strategy:

> At Comdex [trade show] in 1991, Grove delivered a keynote speech . . . and showed how a notebook PC equipped with PCI and special computer chips could receive E-mail messages and graphics delivered over a wireless network. At the time, that was a real breakthrough. . . . He [Grove] realized that . . . leadership could become Intel's key competitive strength. "That was the 'Aha!' for me," he says.
>
> —(Kirkpatrick, 1997, p. 64)

In 1991, Grove created the Intel Architecture Labs in Hillsboro, Oregon, with research focused on the PC. In 1997, it had 600 employees, mostly programmers. Many of the lab's projects involved creating new software. The purpose of the lab was to drive new multimedia applications for PCs, in order to create continuing demand for new generations of higher performance PCs. Intel had taken its technological future in hand, fostering the new applications that will need new PCs, powered by new generations of Intel chips:

> Ron Whittier, Intel's point man on the media lab, is senior vice president in charge of content. He spends all his time figuring out ways to get com-

pelling material—for entertainment, leisure, and business—ready for future generations of PCs.

—(Kirkpatrick, 1997, p. 71)

By 1996, the new research efforts at Intel were beginning to pay off:

Grove's PC efforts have already had a big impact. Michael Slater remembers that the PC of 1991 was "very weak" when compared with Apple's Machintosh. But those differences have narrowed—very much because of Intel's attention to the platform.

—(Kirkpatrick, 1997, p. 72)

Thus, in the 1990s, Intel was beginning to perform the kind of corporate research to make it a technology leader (while Apple was still riding principally on Xerox's research of the 1970s and 1980s).

EVALUATING CORPORATE R&D

Since R&D is an investment in the corporation's future, it should ultimately be evaluated by the return on investment. In practice, however, this is difficult to do because of the time spans involved. There usually is a long time from (1) funding research to (2) successful research to (3) implementing research as technological innovation to (4) accumulating financial returns from technological innovation. The more basic the research the longer the time for it to pay off, and the more developmental, the shorter the time. For example, the times from basic research to technological innovation have historically varied from a minimum of ten years up to 70 years. For applied and developmental research, the time from technological innovation to break-even has been from two to five years.

In addition to the varying time spans, the different purposes of research also complicate the problem. Research is aimed at maintaining existing businesses, beginning new businesses, or maintaining "windows on science." Accordingly, evaluating contributions of R&D to existing businesses requires accounting systems that are activity-based, can project expectations of benefits in the future, and can compare current to projected performance. The evaluation of research needs to be accounted to these purposes of research:

1. R&D projects in support of current business
   - The lifetimes of current products are projected.
   - This product mix is then projected as a sum of profits.
   - The current and proposed R&D projects in support of current business are evaluated in terms of their contribution to extending the lifetimes or improving the sales or lowering costs of the projects.

2. R&D projects for new ventures
   • Projects that result in new ventures are charted over the expected return on investment of the new ventures.
3. R&D projects for exploratory research
   • These projects are not financially evaluated, but treated as an overhead function. They are technically evaluated only on their potential for impact as new technologies.

Many scholars have tried to simplify this logic by trying to create analytical "shortcuts" to the evaluation problem. For example, Richard Foster, Lawrence Linden, Roger Whiteley, and Alan Kantrow suggested trying to trace the following logic as an expression (Foster et al., 1985):

$$\text{return} = \frac{\text{profits}}{\text{R\&D investment}}$$

This can be expanded as:

$$\text{return} = \frac{\text{profits/technical progress}}{\text{technical progress/R\&D investment}}$$

While this expression may be seem appealing, its problem is in measurement. While profits and R&D investment are measured in dollars, there is no general measurement for technical progress across all technologies.

In conclusion, there is no useful shortcut for evaluating R&D aside from accounting for it as contributes to business in particular categories. This requires establishing an appropriate accounting system for the research function. For example, Yutaka Kuwahara and Yasutsugu Takeda described an accounting method for measuring the historical contributions of research to corporate profitability (Kuwahara and Takeda, 1990). In the accounting system, profit from each product must be proportioned according to productive factors, one of which is a profit contribution factor by the research.

SUMMARY

Corporate research activities can be classified by the purposes of supporting current business, establishing new businesses, or exploring new technologies. Profit-gap analysis displays the need for product development.

The inclusion of science into the company framework, through the corporate research laboratory, was a significant event in institutionalizing scientific technology. Organizing industrial research in a mixed form of central research and divisional labs attends to both long-term and short-term technology needs of a firm, but must be managed for cooperation. Formalized procedures are important to promote effective communication, cooperation, and integration of research and business activities.

Since R&D is an investment in the corporation's future, it should be ultimately evaluated by its return on investment. In practice, however, this is difficult to do because of the different time spans involved. Accounting for research requires an accounting system with the capability of analyzing the contribution to profits over the long term from different productive factors, including research.

FOR REFLECTION

In the literature on research and technology, find a study of a corporate research lab and evaluate (qualitatively) its contribution to the corporation. Were new product lines begun from research? Were new businesses begun from research? Was production improvement begun from research?

____17
DESIGNING PRODUCT

CENTRAL CONCEPTS

1. Design process
2. Core and supporting technologies in products
3. Hardware, software, and service design
4. Make-buy design decisions
5. Innovation of new-technology products
6. Market acceptance of product innovations
7. Product development process
8. Product development time
9. Product families
10. Next-generation product platforms

CASE STUDIES

The Personal Digital Assistant Product in the 1990s
ASIC Chip Design Process

INTRODUCTION

Let us now turn to examining, in more detail, the activities of designing new products. All products begin as ideas—inventive ideas of technologies that are embodied into the materials, processes, procedures, or logics of the product. Accordingly, the design of a product is an embodiment of an idea to meet a customer's need. New technology for high-tech products is implemented in the product design stage of engineering. This is true whether a product is a hard good, software, or a service (or any combination of these).

The design phase of new product innovation is a critical stage. For example, studies on many different kinds of products have reached the conclusion that at least 75% of the eventual total cost of a product is determined in the design phase (National Research Council, 1991).

DESIGN OF PRODUCTS

The differences between technology, product, and service are as follows:

- Technology is a *knowledge* of the way to manipulate nature.
- Product is a *designed embodiment* of technologies in a good or process.
- Service is the *application of products* for functional activities.

The logic of design centers around the intellectual dichotomy of function and form. Design is creating form (morphology and logic) to perform function. For example, the design of a piece of hardware, such as a hammer, has the function of driving nails into wood in a customer's construction application. The form of the hammer consists of a heavy metal head and a long wooden or fiberglass handle. The head has a flat surface for driving nails and a clawed surface for extracting nails. The combination of the weight of the head and the length of the handle facilitates the application of force by the customer using the hammer.

All products, whether hardware, software, or services, can be analyzed as to the function and form in their design; these provide the criteria for distinguishing a good design from a bad design. A good design provides an appropriate form for a function. A bad design is one whose form provides a function that does not perform adequately or safely and at the right cost. Designs of products that provide poor performance, are unsafe, and have a high price result in products that do not sell.

The logic of the design process for a product can be divided into several phases:

1. Customer requirements
2. Product specifications
3. Conceptual design
4. Preliminary design
5. Detail design

6. Product prototype
7. Testing
8. Final design

All phases of product design create opportunities and risks. For example, determining who the customer is and what the product requirements are for a customer is a critical judgment. This involves understanding the potential of market niches and application systems in these niches.

Setting the specifications requires translating customer needs to engineering specifications; this procedure is never very clear. In fact, creativity and innovation in this translation often result in higher-quality products than in a more plodding and literal translation. In addition, a given product will be part of a product family in order to cover market niches. Thus, product design will occur within a broader activity of product family strategy. Within this strategy, product architecture and generic product platforms are critical decisions for profitability and competitiveness.

Finally, the management of the product design process itself is important both for innovation and for iterative product designs. Stephen Rosenthal and Anil Khurana sketched some of the logic for structuring the conceptualization phase of new product development as:

- Identifying a product opportunity from the basis of existing product portfolio strategy and market and technology analysis
- Formulating a product concept
- Defining the product and planning a development project (Rosenthal and Khurana, 1997)

Case Study: The Personal Digital Assistant Product in the 1990s

An example of a new product introduction that didn't quite make it was the "personal digital assistant" introduced by Apple in 1993. It involved both hardware and software, and also illustrates the integration of software and hardware in advanced technology projects. The historical setting of the case study is the second decade of the personal computer industry, when leaders were trying to innovate new directions in personal computing. The CEO and chairman of Apple Computer, John Sculley, had taken a personal interest in introducing a new computer product for Apple that would be innovative.

We recall that Sculley had been selected as president for Apple by his predecessor, Steve Jobs and that Sculley subsequently ousted Jobs in a power struggle over leadership in Apple. Sculley, who had come to Apple from a long career in marketing soft drinks, tried to gain the reputation of being a high-tech guru, like his predecessor, and introduced the personal digital assistant (PDA).

By 1996, this product was showing growing pains in the market:

> They were dubbed personal digital assistants, and they were supposed to be the ultimate information appliance. Nifty little handheld devices, PDAs would become ubiquitous tools that would hold telephone numbers, keep your calendar, store notes and send and receive data without wires. At least that was what John Sculley, former chairman of Apple Computer Inc., predicted in 1993 when he took the wraps off Newton, the pioneering PDA.
> —(Elstrom, 1996, p. 110)

As a product, Newton was over-hyped and underwhelming, with sales beginning to shrink in 1995 at 381,000 units, compared to 389,000 units sold in 1994. They were not selling because they were not delivering good enough performance for a market application:

> PDA . . . may hold important lessons as the world's electronics manufacturers attempt to build other information appliances. . . . These hybrid gadgets weren't good in either role they aspired to. They were poor computers and lousy communicators.
> —(Elstrom, 1996, p. 110)

It is important for a new product to find a market application and deliver the features and performance at an appropriate price for the application. This is the bottom line of product design. In the case of the PDA, it had not been designed to provide wireless communication. In 1996, one could buy a wireline modem that let the Newton (or the competing product, the Hewlett-Packard HP200LX) retrieve E-mail, but this was a $300 extra. Adding a modem to work on cellular networks cost $500 extra. The design of the PDA should have seen communications as an essential feature:

> The question is whether the PDA will become a great communicator before wireless smart phones take on all the attributes of a PDA. It doesn't make sense to have them as two separate devices. . . . If I have something that stores telephone numbers, I want it also to be a telephone.
> —(Elstrom, 1996, p. 110)

SYSTEMS ANALYSIS IN DESIGN

Making function and form match in a design requires a systems approach to the design process. The product must be viewed as an engineered system, with a functional logic mapped into a product morphology. Systems analysis of a product design is a systematic depiction of the functional transformations required for product performance. A systems analysis can be partitioned into design focuses on:

- Boundary
- Architecture
- Subsystems
- Components
- Connections
- Structural materials
- Power sources
- Control

CORE AND SUPPORTING TECHNOLOGIES IN A PRODUCT

The core technologies of a product are the technologies that provide the principal transforming capability (function) of the product. Supporting technologies of a product are the technologies that provide needed or desired features in the product for an application.

For example, in an airplane, the function is flight, and the core technologies for this are wing lift, engine power, and aerodynamic controls. For a passenger application, supporting technologies include seating and services for passengers and storage of baggage and freight. For a military application, supporting technologies include weapon and defense systems.

In a computer, the core technologies are in the central processing unit, memory, data buses, and systems control software. Supporting technologies are in input-output and peripheral devices.

In software, core technologies are the procedures and algorithms that enable the primary manipulation capabilities of the software. Supporting technologies are procedures and algorithms that add features and computer system software that provide platforms for operating the software.

The concept of core and supporting technologies also applies to services as a delivered product. For a service, core technologies provide the diagnostic and delivery of the primary function of the service, and supporting technologies add the procedures and services that assist and complete the service delivery. For example, in medical services, the doctor's knowledge, skills, and equipment provide the core technologies for diagnosing and treating illness; the nurse's knowledge, skills, and equipment provide the supporting technologies for caring for the patient and completing a prescribed course of treatment.

DESIGN OF HARDWARE OR SOFTWARE

From a technology perspective, the differences between hardware products, software products, and services lie in the relative complexity of the morphology or schematic logic of the core technologies of the product:

- In hardware design, morphology is complex and schematic logic simple.
- In software design, morphology is simple and schematic logic complex.
- In service design, both morphology and schematic logic are complex.

The reason that managing hardware design projects has been better understood than managing software design projects is that engineers are generally better trained in physical morphology than in logic (see Table 17.1).

A mechanical engineer receives training in mechanics and thermodynamics. An electrical engineer is trained in electricity, magnetism, and computer sciences. A chemical engineer's training is in chemistry and, sometimes, biology. A civil engineer receives training in mechanics and hydrology. A computer engineer trains in electromagnetic physics. Only the computer scientist receives training in some schematic logic.

Most engineers receive only limited training in schematic logic. For the engineer, the training in logic is usually limited to the mathematics of deductive and inductive syntaxes (e.g., algebra, calculus, differential equations, matrices and vectors, and probability and statistics). The only transforming logics in which an engineer receives training are the generic functional logics of the product designs. For example, the mechanical engineer is familiar with the spatial logic of geometry, the electrical engineer with the transforming logics of generic circuits, the chemical engineer with the transforming logics of generic chemical engineering processes, and the computer electrical engineer with the circuit logics of computational systems.

Thus, the physical aspect of product design is usually better understood by engineers than the schematic logic aspect of product design (see Table 17.2), except for hardware products.

Deductive/inductive syntaxes are logics that include the topics of mathematics that enable science and engineering to create quantitative models and simulations of physical phenomena and the physical morphologies of product operations. Mathematics used to create deductive and predictive models include algebra, calculus, differential equations, matrices, and vectors. Mathematics used to create inductive estimates of product properties include probability and statistics.

TABLE 17.1 Science Bases for Physical Morphology

Mechanical physics
Electromagnetic physics
Chemistry
Materials science
Environmental sciences
Biology
Computer sciences
Mathematics

TABLE 17.2   Schematic Logics

- Deductive/inductive syntaxes
- Transformation procedures
- Decision algorithms
- Control system techniques
- Application languages

Transformation procedures are logics that arrange for the transformation of inputs (such as data, signals, or resources) to outputs (such as information, processed signals, or products).

Decision algorithms are logics that systematically structure decisions and seek optimal solutions.

Application languages are software tools that facilitate generic activities and procedures (e.g., word-processing, accounting, and data bases).

HARDWARE DESIGN

Hard goods are physical products that can be used to satisfy physical needs, transform materials, carry out activities, or perform services. All hard goods are embodied in material manifestations, requiring materials and power resources. The physical aspects of hard goods require geometric and material design and manufacturing techniques.

The schematic logic of hard goods is in the control subsystem of the product. Traditionally, technologies for products and technologies for services had been relatively distinct; but with modern computer and communication technologies, they are increasingly sharing technologies for information and control in both product and service systems. Modern products now often embed control technologies in hardware and software.

Technical innovation can be embedded in a product/service design in either (1) new physical forms, materials, or power sources of the product, or (2) new control systems in schematic logic in the design of the product's operations.

SOFTWARE DESIGN

In contrast to hardware design, software design is principally concerned with complex schematic logics. Software design is the creation of schematic logics for a specific application. The elements of the schematic logic in software are conceptual linguistic primitives, such as "nominal terms (names) and "relational terms" (operations).

Thus, it is important to software design to understand the nature of language. Language is the basic tool of thinking. We think in language, involving either an internal dialogue using language when we think to ourselves or an external dialogue using language when we think with other people or computers. Thinking with other people or computers is usually called "communication" or "information." We learn to think in language as we interact with our parents as young children and acquire language. (Rare but poignant studies of children who grew up in isolation from human interaction have shown that language acquisition and thinking ability are interactive and must be acquired when very young.)

What language principally does is:

1. Sharpen and refine perception
2. Abstract and generalize experience from one specific event and context to another
3. Facilitate social cooperation and conflict

Therefore, the development of language affects the nature of perception (how and what we see in the world), the nature of thinking (how and what we abstract and generalize of the world), and how we interact with one another.

Software design is the development of specialized languages and linguistic tools to facilitate thinking, computation, information, communication, cooperation, and competition.

A language is composed of a set of nominal and relational terms and a grammatical structure. The set of nominal and relational terms of a language constitutes the "dictionary" of the language. The grammatical structure of the language constitutes the architecture of the language.

A linguistic logic is a kind of language to talk about the architecture (grammatical structure) of a language. Linguistic logic is a one-step linguistic regression of a language. As one constructs a linguistic logic as a language about language (first linguistic regression), one can also construct a language about a linguistic logic (second linguistic regression); this is often called a "meta-logic." Software design can thus be regressed into the design of the software language, software logic, and software meta-logic. This is why we emphasized earlier that software design is complex in logic but simple in physical morphology.

The software language is what is coded in the software development. The software logic is the architecture of the software development. The software meta-logic is the boundary and assumptions about the software architecture.

In summary, in designing software, coding tasks must include the expression of:

- *System architecture*—for example, a word processing program has an architecture of sentences, paragraphs, pages, and documents.

- *Primitive linguistic operations*—for example, in a word processing program, linguistic primitives express fundamental editing operations, such as delete, insert, erase, move, and spell-check.
- *Data input and output*—for example, in a word processing program, features must provide for inputs and/or outputs from several sources, such as keyboard, storage disks, optical scanning, and electronic transmission.
- *Coordination of activities*—for example, in a word processing program, precedence ordering allows some operations to be called only after other operations; e.g., one must open a file before typing into it.

Software design, therefore, begins with a system approach to charting the kinds of allowed information and operations and possible flows among kinds of information and/or operations. Teams of programmers cooperate to write sections of the code and integrate the sections into an overall program. Debugging is a critical activity of software production.

Quality problems in software divide into bugs and defective disks or transmission. Failure of software to run properly is called a "software bug." Failure of software to run from a particular storage medium, disk, or transmission is called a hardware failure. Software bugs are the most serious problem for a software producer.

Software bugs arise from several sources:

- Poor programming
- Complexity of the application
- Newness of the application

The quality of programming depends on the skills of the programming team. The complexity of an application determines the complication of the architecture, coordination, and number of lines of code. The more complex the application, the more bugs will occur simply from the inability of a team to comprehend the totality of the program operation in application. The more innovative the application, the less experience will be available to determine the scope and details of the application. Thus, bugs will arise from users trying to do something in the application for which programming was not planned.

Because of these sources of problems, the rule of thumb for large software programs is that bugs will never be entirely eliminated. Software producers must, therefore, depend on determining quality by the *rate of the occurrence* of bugs and not the absolute number of bugs. For this reason, software producers need to have marketing policies to quickly upgrade software versions and replace older versions when bugs are found by customers.

Accordingly, technological innovations that provide new functionality, extend functionality over more applications, or find bugs faster provide powerful competitive edges to software producers.

SERVICE DESIGN

Services are activities that provide value-added transformations, transactions, communications for a customer. Examples of service industries include banking, rental properties, medical services, accounting services, advertising services, retail services, and food preparation services. As a sector of the economy, services have been growing to provide the majority of areas of employment for a developed economy.

Service technologies are the tools and procedures used in the development and/or delivery of services.

Services can be internal to a productive organization or external, as one of the products of the organization. Internal services provide activities necessary to the productive operations of the organization, such as engineering, personnel, marketing, manufacturing, and finance. External services are the products sold to customers or the assistance provided to customers.

Innovation of a new external service is the creation of a set of activities that can be sold to customers. Innovations in service technologies are frequently dependent on either new hardware or software. New schematic logics are important sources of innovation in service technologies.

Technologies for the service industries (or for internal services within a firm) use products and devices from manufacturers, but in a procedural system that requires information and communications. Therefore, what is unique in a service application is the information and communication and control procedures. The design of software for information handling, transactions, communication, and control is essential to services.

MAKE-BUY DESIGN DECISIONS

Not all parts of a product will usually be manufactured by a business; some will be bought. In the design of the product, an important decision is which to parts of a product to make and which parts to buy.

This same decision applies to products that are hard goods, software, or services. In software, for example, a critical "buy" kind of decision is what operating system to program the software on which to run. In service delivery, there can be decisions on what secondary services to provide or to contract. For example, in the 1990s in the personal computer industry, the dominance of Microsoft operating systems meant that far fewer programs were being written to run on Apple computers. As another example, in medical practice, general medical practitioners refer patients to medical specialists for specialized treatments or surgery.

Criteria for the make-buy decision are competitiveness, timing, cost, quality, and safety. The most important components of a product to be made are those that provide a competitive edge in the product for the firm, such as the components that embody the core technologies of the product. However, if some of the core technologies require a substantially different expertise, production capability, or capital investment, then an industry might purchase some of its core technologies.

For example, in airplanes, two core technologies are airframe and engine. Airplane manufacturers make their airframes but purchase their engines. Airplane safety and cost are the primary competitive factors. Since safety requirements prescribe more than one engine on commercial aircraft, airplane manufacturers can purchase the engine and design the plane to enable it to fly on one engine alone. Engine technologies are so different from airframe and airplane control technologies that the airplane manufacturers specialize in the latter two.

If all components are purchased, then the business has no competitive technology edge in its product and no way to differentiate product. On the other hand, if all components are made, then the business will have to run several businesses competitively to produce one product. The skills and plants involved, the capital required, and the relative profit margins for the different components may make this economically inefficient.

For example, after the Second World War, the U.S. automobile industry produced up to 75% of the components of an automobile (the major exception was tires). After increased worldwide competition in automobiles grew in the 1970s and 1980s, some U.S. automobile manufacturers increased the purchase of components up to 75%.

In innovating a new product, the purchase of components speeds up the time of design and development to get the product to market. However, with the purchase of components, the company loses control over the ability to differentiate the product and control costs and quality.

After the decisions to purchase materials or parts for a new product, the quality and performance of the purchased supplies affect the quality and performance of the new product. How one manages the supply-vendor relationship is critical to the quality of the product.

In the design process, it is important to include potential vendors in a concurrent design process, utilizing their expertise to gain higher material or part performance with lowest cost.

INNOVATION OF NEW-TECHNOLOGY PRODUCTS

If a product design is to implement new technology, then many problems are important in the design conceptualization phase, such as:

- Determining the best application focus for a new-technology product or service
- Determining appropriate requirements and specifications for a new-technology product or service
- Making the appropriate performance/cost tradeoff in designing the new-technology product or service
- Deciding where the proprietary and competitive advantages of a firm's product should lie

Product design requirements are determined by the class of customers and the applications for which the customers will use the product. In technology innovation, the most easily approached first market is an established market in which the new technology can substitute, since some knowledge of customer applications and product requirements already exists for that market. Products or services with embedded new functionality create brand-new markets; improved performance products or services of existing functionality substitute in existing markets. For radically new products, the first critical design decision is identifying the customer's application for the product, and a substitution market can provide guidelines.

The activity of designing a product requires understanding the technologies that will be embedded in the product and decisions on the tradeoff between desirable aspects of a product design, constraints of economics, and limitations of current technology. Usually, the cost of an innovative product will be higher than the price of existing products for which it is to substitute. A second critical design decision is estimating the acceptable sales price to a customer of the product and a required production cost for profitability at such a sales price.

Different markets for new technology may value the performance/cost tradeoff in product design differently:

1. In the military market, performance will usually be valued over cost.
2. In the industrial market, performance will usually be valued at a cost that can be justified as an appropriate return on investment.
3. In the business market, performance will usually be valued within the upper boundaries of the business practices of monthly expenses (leasing will be used for large-ticket items).
4. In the hobbyist market, performance will usually be valued within the upper boundary of disposable recreational income.

5. In the consumer market, performance will usually be valued within the price ranges of comparable categories of consumer purchases.

Therefore, for each market, the design performance/cost tradeoff must be made differently. Innovative product designs can, therefore, fail commercially several ways:

1. The product design may fail if its expression of a principal technology system performs functionally less well than a competing product.
2. The product design may fail in its balance of performance and features as perceived by the customer (even if technologically performing as well as a competing product).
3. The product design may fail if it is priced too high for the price ranges of an intended class of customer.

DYNAMICS OF HIGH-TECH PRODUCTS

As technology progresses, the products in which the technology is embedded will substitute for existing applications and start new applications. The product requirements and specifications are determined by the applications; as the application systems develop, then product requirements change. This is a kind of chicken-and-egg problem. Which comes first in technical progress, product specifications or application system requirements? The answer is both, since they are interactive.

When radically-new-technology products are innovated, the immediate applications that are found for them soon indicate the performance limits of the technology. This provides incentive to improve the performance of the technology guided by the applications. As performance in the product improves and costs decline, then new applications open up for the product.

In a new technology, products often go through several generations of next-generation technology products. Yet, in each generation, there will be tradeoffs between desired performance and cost. As technology progresses, product design must focus on optimizing product performance within the constraints of currently limiting technological capabilities and economics.

Case Study: ASIC Chip Design Process

Antonio Bailetti, John R. Callahan, and Pat DiPietro studied a product design process in an application-specific integrated circuit (ASIC) chip design department of a telecommunications firm (Bailetti et al., 1994). They looked at three chip design projects, all three producing specialized chips for telecommunications switching functions, and examined problems that occurred in the design projects. They diagrammed the information and decision flow interactions of the design process, as depicted in Figure 17.1.

The figure indicates that a program manager and a project manager interacted on setting both the project schedule and the technical overview of the project. The technical overview was shared with manufacturing to ensure that the product design could be manufactured with an existing manufacturing system. Manufacturing drew upon a knowledge of process technology that included a computerized design and process aid called a "cell library," which was managed by a cell library group.

The project manager set and monitored a chip design schedule, which had a chip manager for the design, who supervised a design team and layout specialist. The design team, in cooperation with the layout specialist, produced a chip design that embodied the design stages of creating a "high-level design," a detailed design "schematic," and "netlist." The chip design was simulated and modified according to simulation results. Electronic chip design in the 1990s used very sophisticated sets of computer-aided design (CAD) and computer-aided engineering (CAE) software tools.

This case shows that a product development process requires an interaction between design teams and manufacturing representatives, coordinated in a product development process, with program, project, and product development managers. Bailetti, Callahan, and DiPietro emphasized

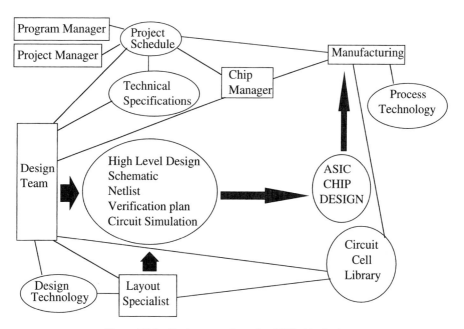

Figure 17.1. Design procedures for ASIC chip design.

the importance of coordination across the different groups and within the teams. They found, in the three projects, that costly delays in design were due to poor coordination. For example, in one project, an error in the cell library resources caused a serious design problem that cost the firm 16 weeks' delay and, indirectly, $450,000. Vendors had given assurance about information in the library which, in fact, was not there. The group responsible for the cell library resources should have picked this up if it had been aware of the importance of the information to the design project.

PRODUCT DEVELOPMENT PROCESS

The product development process requires cooperation within a design team that coordinates across the functions needed for good product development. Accordingly, the product development team should consist of representatives from the different relevant activities of:

- Research
- Product engineering
- Manufacturing engineering
- Marketing and sales
- Finance

The job of the team is to formulate the design requirements and specifications, to initially take into account considerations of manufacturing, marketing, finance, and research into the product design as early as possible. The team should conduct a competitive benchmarking of competing products in all price categories and establish a list of best-of-breed performance and features for the product system. It also needs to have a product development schedule and early prototyping goals and means. It needs to identify early sources of supplies, and draw on their expertise and suggestions about part and subassembly design. The team also needs to consult and solicit suggestions from customers, retailers, service firms that repair and maintain the product systems, and insurers for suggestions for product improvement and feature desirability. The team manages the product development process to encourage teamwork and cooperation in developing a product rapidly and of the highest attainable quality. It is also important for the team to interact with and exchange information with other development projects in a firm. For example, Kentaro Nobeoka and Michael A. Cusumano assert that there are often important interactions between simultaneous development projects, as, for example, when developments from one ongoing project are transferred to another, parallel, ongoing project (Nobeoka and Cusumano, 1995).

Modern product development processes make extensive use of modern design systems.

DESIGN SYSTEMS

Computer-aided design systems have become critical to the product development process. For product designs, design systems that provide geometric representations, or CAD (computer-aided design) systems have become essential to modern product design. Added into CAD systems is the ability to test performance, part of the design system called CAE (computer-aided engineering). The most sophisticated CAE systems in the 1990s were in electrical engineering, and the most sophisticated CAD systems were in mechanical engineering. The most sophisticated CAD and CAE system combinations were in aeronautical engineering.

In the 1990s, most research to improve engineering design was being incorporated into CAD and CAE systems. The goal of engineering design for products was to design the product and test the product virtually before a physical prototype was created. (In the 1980s, the first commercial airplane to be designed entirely in CAD and CAE systems without first producing a physical prototype was the Boeing 777.)

The use of computer-aided design systems was also changing the product development process. In the mid-1990s, computer-aided design tools and concurrent engineering practices constituted an area of advancing technology in their own right. Nelson King and Ann Majchrzak studied several groups of designers using CAD and CAE systems within concurrent engineering practices, and noted that the assumptions of how designers would use these tools and how practice actually occurred were markedly different. In particular, they found that there were still many technical problems, so that fully integrated computerized design systems were not yet available for concurrent engineering practices and developers of such systems were making naive assumptions about the kinds of information designers needed and how they used information. They concluded that as long as the technology continued to change:

> [A]ssumptions made by the concurrent engineering tool development community are likely to inhibit [these] tools from [full] success. . . .
> —(King and Majchrzak, 1996)

However, despite the problems, modern design activities are already critically dependent on computerized design systems.

PRODUCT DEVELOPMENT TIME

Development time is another important factor in the commercial success of new products. As we noted earlier, being too late into a market after competitors enter first is a disadvantage, unless one comes in with a superior product or a substantially lower price. Critical to fast product development times is the use of multifunctional product development teams. B. J. Zirger and Janet L. Hartley studied the product development times of several electronics companies (Zirger and Hartley, 1996). They found the fast product developers had cross-functional development teams that included explicit goals about fast time to market and overlapped their development activities in concurrent engineering practices.

Jeffrey L. Funk studied product development at Mitsubishi's Semiconductor Equipment Department and at Yokogawa Electric. He summarized that studies about the product development practices of Japanese firms identified several strengths in product development, including: (1) the use of multifunctional teams in problem solving, (2) close relationships with customers and with suppliers, (3) focus on incremental improvements to product and production, and (4) learning how to improve the product development process (Funk, 1993).

While product development time is important, it is, of course, not the only factor necessary for a successful product. For example, M. H. Meyer and J. M. Utterback examined whether increased product development time alone guaranteed commercial success and, not surprisingly, found that it did not (Meyer and Utterback, 1995). Commercial success for new products depends on many factors, of which time to market is only one. There are many factors that may increase development time in innovation, such as the complexity and number of new technologies that have to be integrated into the product. The minimum condition for a good chance at success is to get the product performance right at the right price for the right application for a customer. The sooner this can be done, the better.

In software development, the complexity of the logic and the production process particularly interact, creating special challenges for software development processes. Robert B. Rowen emphasized that technical difficulty in large software projects arises from both the complexity of the application and the process of a large team of programmers (Rowen, 1990). He argued that these two factors stimulated the traditional logical sequence of large-systems design activities:

- System feasibility
- System requirements
- System architecture design
- Detail component design
- Component code production
- System code integration
- System testing
- System implementation
- System maintenance

Rowen further maintained that prototyping of software as these stages advance improves software production by creating iterations in the stages of this activity.

PRODUCT FAMILIES

Products can be varied to design a product family to cover the niches in a business's market. Variation of a product into different models of a product family is a redesign problem, altering the needs and specifications of the product model to improve focus on a niche of the market. A product model may be replaced by an improved or redesigned model. The time from the introduction into the market of a product model and its replacement by a newer model is called a "product model lifetime."

Product models in a family may be replaced by a newer set of product models; this is called a new generation of the product. Product generations are designed to provide substantial improvement in performance, features, or safety and/or substantially reduce product cost. The time from the introduction of one generation of a product family until its replacement in the market with a new generation is called the "product generation lifetime."

Steven Wheelwright and W. Earl Sasser have emphasized the importance of long-term product planning—mapping out the evolution of products in a product line (Wheelwright and Sasser, 1989). They pointed out that products often evolve from a core product, such as branching out enhanced models for a higher-priced line or stripping a model of some features for a price-reduced line. They also argued that from the core product, one should plan its evolution in generations of products. The core product system will express the generic technology system, and higher- or lower-priced versions will differ in the subsidiary technologies of features. Next-generation products differ in dramatically improved performance or new features, or improved technologies in features or dramatically reduced cost.

One can diagram the planning of product lines as branched products and as generations of products; Wheelwright and Sasser called this a "new product development map." Such a map is particularly useful for the long-term product development process, when the anticipated technical goals for product performance and features are listed in the product boxes. It is also helpful to summarize needed technology developments against new product ideas to identify where concurrent engineering cooperation between research and product design and manufacturing should be planned.

Product family planning is especially important to deal with competitive conditions of shortened product life cycles. Short product life cycles can decrease profits. For example, Christoph-Friedrich von Braun charted shortening product lifetimes from 1975 to 1985 in Siemens, with an average of 25% shortening (von Braun, 1990 and 1991). With shorter lifetimes usually comes smaller cumulative product volume sold, and, hence, less return on investment from the R&D to introduce the product. This makes the idea of the product family important by extending product lifetimes over a variety of product models and product model improvements.

PRODUCT FAMILY PLATFORMS

From the perspective of technology, the key to efficiently designing and producing a product family is to develop common technology platforms for the core technologies of the product. Common product platforms that can be modularized can efficiently be varied to adapt the product line to different market niches.

For example, Susan Sanderson and Mustafa Uzumeri studied the successful innovation and subsequent market dominance of the Sony Walkman product line, concluding that:

> Success in fast cycle industries (e.g., consumer electronics) can depend both on rapid model replacement and on model longevity. Sony was as fast as any of its chief competitors in getting new models to market, an important explanation for the wide variety of models . . . is the greater longevity of its key models.
>
> —(Sanderson and Uzumeri, 1995b)

The key models provided common technology platforms from which to vary models. In general, for any product family, Sanderson and Uzumeri have stressed the importance of developing several generations of product platforms for successive generations of product families, used to (1) vary models for covering market niches with a full product line and to (2) advance the performance of generations of product lines to stay ahead of competitors (Sanderson and Uzumeri, 1995b).

Marc H. Meyer and James M. Utterback also have discussed the concept of product platforms to provide the basis for individual product design (Meyer and Utterback, 1993). They also stressed the importance of innovating new technical generations of product platforms to maintain the competitiveness of product families. The core technologies underlying a product platform provide the opportunities of next-generation platform advances.

Thus, the technology key to keeping product families competitive are next-technology generations of product platforms for product families.

SUMMARY

Designing a product is the stage in which technologies are incorporated and in which innovation for new products or improved performance is applied. The underlying principle of design is to create forms that make function possible. Design requires a systems approach to conceptualize and detail the functional totality of a product.

Core technologies of a product are the technologies that provide the principal transforming capability (function) of the product. Supporting technologies of a product are the technologies that provide needed or desired features for complete a product for an application. Technological innovation in new products requires determining the best application focus, appropriate product specifications, performance/price tradeoffs, and proprietary advantage. The dynamics of new high-tech markets require frequent redesign of product. The more innovative the product, the more likely new markets will be created.

Decisions to make or buy which components of a product strongly affect the competitiveness of the product. The more components are purchased, the shorter the development of a new product design, but the control is lost over product differentiation, cost, quality, and safety. Hardware, software, and service product designs differ in their relative complexity on morphologies and schematic logics.

The product development process requires multifunctional design teams. Product families provide the ability for a product to capture a larger percentage of a market. Product families can be kept competitive over time through next-technology generations of product platforms.

FOR REFLECTION

Identify a product line with a history of product models. Describe the product, market, and product family evolution in the history. Which companies succeeded in which market niches of the product line? What was the critical performance/feature/price balance that contributed to market successes?

____18
PRODUCING PRODUCT

CENTRAL CONCEPTS

1. Economies of scale or scope
2. Production system
3. Unit production processes
4. Production as sociotechnical systems
5. Production technology audit
6. Production quality
7. Production learning curve

CASE STUDIES

Biotechnology Process Innovation at Bristol-Meyers Squibb
Flexible Manufacturing Systems Installation
Production Systems of the Airline Industry

INTRODUCTION

After new high-tech products are designed, they next must be produced in volume before sales can begin. We will now look in detail at the role of technological innovation in production. The goal of technological innovation in production is to produce a product or service at higher quality and lower cost than competitors.

Businesses carry on repetitive economic activities that produce and sell volumes of products to a market. Production is the organizational process that creates product volume; this has been called by different names in different industries. For example, although in the chemical industry this has traditionally been called "production," it is called "manufacturing" in hardware industries. In the civil structures industry, it is called "construction," whereas, in the service industries, it is called "service delivery." We will use the term "production" as a generic term. In this chapter, we will review the forms of production in products and services, and discuss issues about technological innovation in production.

PRODUCTION LOGIC

The logic of any production consists of a "production system" and generic "unit production processes." A production system is an ordered set of activities that transforms material, energy, and information resources into products. The logic of a production system differs according to whether the product is a hard good, software, or service. Generic unit production processes are discrete steps in the transformations needed to produce a product from resources. Unit processes differ by industry.

Production of products and of services differ markedly in their details, yet all production can be logically conceived of as a system of production composed of unit processes.

Case Study: Process Innovation at Bristol-Meyers Squibb

This case study illustrates a research procedure for innovating new production processes. The historical setting is in the biotechnology industry of the 1990s, as new therapeutic protein products were being innovated. Each new product required a new unit production process, as cells had to be genetically modified to produce the particular product. The biotechnology industry had to innovate a new unit production process for each new therapeutic protein product.

In 1994, Barbara Thorne presented an overview of the early phases of process development at Bristol-Meyers Squibb Co., describing the general stages in biotechnology drug innovation (particularly for creating drugs based on antibodies):

- Identify the therapeutic target.
- Obtain and characterize the therapeutic protein (such as an antibody).
- Generate recombinant protein (e.g., antibody) in a recombinant engineered host cell production.
- Evaluate recombinant antibody as a drug.
    - —Suitability of antibody for curative target
    - —Side effects of humanization, such as the loss of affinity as an

> antibody of the recombinant DNA engineered antibody or harmful effects

- Re-engineer and re-evaluate.
    —Redesign recombinant engineered antibody for affinity and safety
- Develop into a therapeutic product (quantity and purity). (Thorne, 1997)

Since recombinant engineered cells produce the desired therapeutic protein (such as an antibody) in tiny quantities, the first problem in drug discovery is to produce enough of the engineered protein for therapeutic testing for effectiveness and safety. And, since each engineered protein must be produced by modifying a group of cells, the speed of development of the production process (batches of modified cells) is critical to the speed of the product development. All new drugs require extensive testing in the United States for effectiveness and safety before being allowed on the market by the federal government.

For this reason, biotechnology-based drug firms have to perform process research, as well as product research. Thorne sketched how the two research procedures, for product and for production, interact on a time line, as shown in Figure 18.1.

The time line of new product development requires two sides: product development and process development.

The time line first begins on the product development side with the choosing of a candidate protein (such as an antibody) targeted for a therapeutic application. Next, a DNA construct must be made which, when inserted in a cell, will produce the target protein. This construct is then handed from the product research side of the innovation procedure to the process research side.

Process research then takes the DNA construct and puts it into a host cell (transfection). The host cell is a mammalian cell line, which then produces and secretes the targeted protein. Both quantity and quality of the DNA constructed protein produced are goals of the process research. When the DNA is inserted into host cells, the DNA may randomly be incorporated in the cell's genome. Some of these locations will affect the survivability and growth of the cell line and the productivity of the cell line in producing and secreting the target protein. Process research then uses techniques for selecting and screening for productive cell lines for the target protein. Once this is accomplished, an initial production of the protein is produced and purified for handing back to the product research side.

The product research side must next begin testing the protein for efficacy. The new product and new processes are then transferred to development for drug testing to meet federal standards. The speed with which the

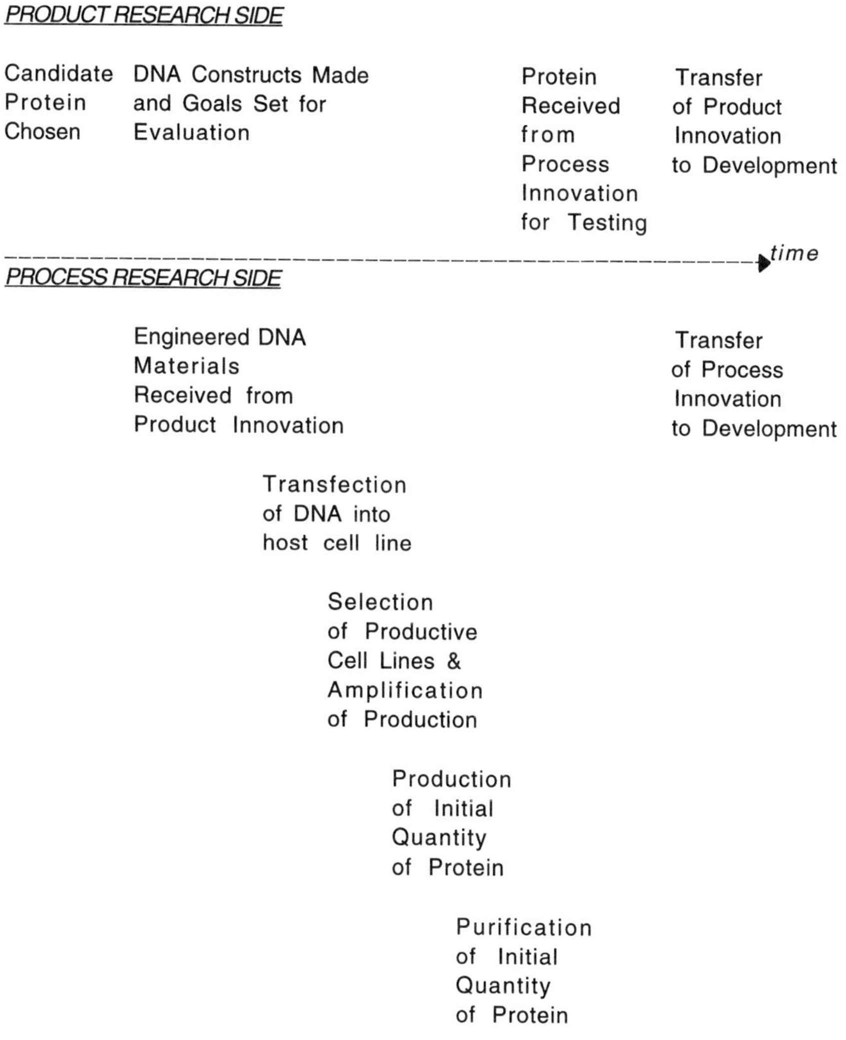

<u>*PRODUCT RESEARCH SIDE*</u>

Candidate DNA Constructs Made   Protein   Transfer
Protein   and Goals Set for   Received   of Product
Chosen   Evaluation   from   Innovation
                                        Process   to Development
                                        Innovation
                                        for Testing

---→ *time*

<u>*PROCESS RESEARCH SIDE*</u>

          Engineered DNA   Transfer
          Materials   of Process
          Received from   Innovation
          Product Innovation   to Development

                    Transfection
                    of DNA into
                    host cell line

                         Selection
                         of Productive
                         Cell Lines &
                         Amplification
                         of Production

                              Production
                              of Initial
                              Quantity
                              of Protein

                                   Purification
                                   of Initial
                                   Quantity
                                   of Protein

Figure 18.1. Product and production innovation time line.

initial production process is created affects the product research side, since it must depend on process research to produce sufficient quantities to begin drug testing.

ECONOMIES OF SCALE OR SCOPE

Innovation in production systems can provide competitive edges of two economic kinds, called economies of scale or economies of scope. An economy of scale is a unit-cost advantage in production that arises from technical efficiencies in production scale. For example, in the materials processing industries (such as steel or chemicals), a larger scale of production can often produce a unit volume of material with lower energy costs and lower waste. An economy of scope is a production flexibility that allows a producer to market a broad range of products. For example, in food processing or retailing industries, an ability to distribute a wider variety of products often creates a larger volume of total sales than smaller competitors can achieve.

When innovating new technology products, economies of scale are important for building distribution capability, creating brand recognition, and establishing a dominant market position. When competitors' production capacities begin to saturate a market and product technologies mature, economies of scope are important to maintain market dominance and maximize returns on investment in production capability.

Technological innovations in production capability can be classified as creating economies of scale or of scope. In the 1980s, for example, innovation for flexible manufacturing used computers for control of machining processes to provide economies of scope over fixed automation production processes. Research continued in the 1990s on flexible manufacturing to create economies of scale without losing economies of scope. Previously, most production research had been for economies of scale; but in the 1990s, as markets became crowded, production research for economies of scope became as important as production research for economies of scale.

For example, Fernando F. Suarez, Michael A. Cusumano, and Charles H. Fine studied flexible production in the printed circuit board industry and observed that manufacturing flexibility can provide flexibility either in product mix or in volume production (Suarez et al., 1995).

The previous case study illustrated the need for rapid process development to complement new product development for the biotechnology products of a pharmaceutical firm. The ability to innovate a variety of new products for a biotechnology company created an economy of scope for the firm; and initial process research was aimed at economy of scope. Later, after a new biotechnology product was innovated, further process innovation would be aimed at reducing the unit cost of production, creating an economy of scale for the firm.

HARD GOOD PRODUCTION SYSTEMS

As noted earlier, production systems are composed of coordinated sets of unit processes. Unit production processes are the discrete transformation steps in making purchased material into fabricated and assembled products. Material unit production processes include:

- Forming processes (such as molding, stamping, machining, forging, reacting, fermenting, lithography, etc.)
- Separation processes (such as distillation, shredding, combing, slurrying, masking, layering, etc.)
- Finishing processes (such as grinding, polishing, painting, coating, filtering, etc.)
- Assembly processes (such as joining, welding, gluing, assembly, fabrication, construction, mixing, etc.)
- Packaging processes (such as boxing, coating, packaging, wrapping, etc.)

Each material unit process embodies a core technology system. Improvement in the technical capability of a material unit production process can come from improved understanding of the phenomena of the technology or from control of the process.

In addition to the set of unit processes, materials need to be moved into and between the unit processes for processing in a production system until the final material parts are assembled into the product. In the chemical industry, chemical materials are moved through pipes. In the mechanical industry, materials are transported by conveyer belts or carts. In the electronic industry, materials are transported by belts, carts, or robots. In the construction industry, materials are transported by trucks, cranes, and people.

Materials must thus be transported to the production site, moved to and between unit production processes, packaged, stored in finished-product inventory, and then shipped into distribution channels. The system of coordinating all this is called the "production system." Control in the production system operates at two levels: scheduling of production and control of work at and between unit processes.

UNIT PROCESSES FOR HARD GOODS BY INDUSTRY

Material unit processes differ by industry. Chemical production uses a set of unit processes, such as reacting, distillation, and filtering. Mechanical production uses a different set of unit processes, such as swaging, milling, machining, molding, forging, assembly, and painting. The set used by the electronic industry includes wafer growth, lithography, etching, metalization, component insertion, soldering, and assembly. Construction uses another set, such as pouring, footing, framing, erection, siding, roofing, insulation, wiring, and plumbing.

SERVICE DELIVERY SYSTEMS

In contrast to hard good production systems, the logic of service delivery consists of other stages:

1. *Service referral*—A customer must arrange for a service by contacting the service deliverer, as, for example, selecting a doctor or opening a bank account.
2. *Service transaction*—Service delivery must be scheduled, as, for instance, making an appointment and visiting the doctor's office or writing a check or making a deposit or withdrawal from the bank account.
3. *Selection of service application*—The appropriate application of the service must be selected, as, for example, diagnosis by a doctor of the patient's illness or recording and accounting a banking fund transaction.
4. *Service application*—The selected application in the service must be provided, such as prescribing a drug or performing surgery, or transferring of banked funds electronically or in cash.
5. *Payment for service*—Payment must be received by the service provider from the customer, as, for instance, billing a patient's insurance company or billing a client's bank account.

Technologies are used in the different stages of the service delivery, such as devices and techniques, information and communication technologies, and professional knowledge base technologies.

PRODUCTION ORGANIZATIONS AS SOCIOTECHNICAL SYSTEMS

Because people, as well as equipment and tools, are organized together for production, a production system can be viewed as a "sociotechnical" system. A sociotechnical system is a concept that emphasizes that the totality of an organized system is a set of human activities organized around a set of technical procedures. The different logics of the production system for hard goods, software, and services create different sociotechnical systems for the firms in these different industries.

In a hard-good production firm, the technical procedures are determined by the material transformations necessary to produce a hard good. The activities of people in the sociotechnical system are also determined by operating the material transformations. Accordingly, production in the different hard-good industries creates cultural differences between them due to their sociotechnical character. For example, the chemical industry performs "continuous" processing in production, whereas mechanical product industries perform "batch-type" processing in production. The building industry usually constructs on-site, whereas other industries produce in factories. Also, we recall that hard-good productive organizational cultures can be divided into "mechanistic" cultures and "organic" cultures, depending on whether the production is high volume of a standard product or low volume of different products. This kind of dichotomy further distinguishes hard-good producing cultures.

Since much hard-good part production is performed in small businesses, an especially important issue is the willingness of a smaller firm to adopt new production technology. Jean Harvey, Louis A. Lefebvre and Elizabeth Lefebvre studied a sample of small manufacturing firms (Harvey et al., 1992). They concluded that companies with tight management of capacity, good capability of process design, well-trained employees, flexible production systems, and good labor relations were the most likely to be the first to adopt new manufacturing technology.

In service delivery, technical procedures are determined by the routinization or customization of the diagnostic and application stages of the service delivery. If diagnosis and application can be completely routinized and are low-risk (as in the airline industry), then most decision-making activities can be programmed and most contact with customers will be provided by relatively low-skilled personnel. In contrast, if the diagnosis and application are mostly customized and high-risk (as in the medical industry), then decision-making activities will be provided by a highly trained professional and most contact with customers will be provided by professionals (doctors and nurses).

Case Study: Flexible Manufacturing Systems Installation

This case study illustrates how innovation in a production system requires a systematic procedure to coordinate the innovation. The historical setting is in the 1980s and 1990s, when flexible manufacturing processes were being installed in manufacturing companies. Anand S. Kunnathur and P. S. Sundararaghavan proposed proper procedures for innovating new manufacturing systems that utilize flexible manufacturing systems (FMS) technologies, and prescribed a logic for such production innovation. Figure 18.2 illustrates such a process (modified from Kunnathur and Sundararaghavan, 1992).

A *production strategic plan* for introducing a new production system should first be formulated, detailing the goals of production innovation for competitive positioning and setting targets for production cost, capacity and throughput, and safety and environmental goals. From these strategic goals, a complete set of innovation projects is formulated within a *master innovation plan* that lists innovation projects, prioritizes their introduction, allocates resources for innovation, and formulates new manufacturing strategy. *Personnel and finance* necessary for the master innova-

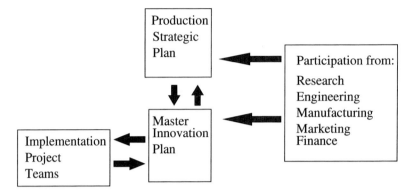

Figure 18.2. Procedure for innovation of flexible manufacturing unit process.

tion plan are trained, assigned, and allocated. *Implementation project teams* of manufacturing, marketing, product design, and information system personnel are formulated for each project. Each innovation project next requires a project plan and management. Finally, *progress in innovating* FMS production processes is monitored and used to modifying and updating the production strategic plan.

Kunnathur and Sundararaghavan stressed the importance of building a manufacturing strategy and a master plan for production improvement, with appropriate input from all business functional areas. They specifically recommended against considering automation (particularly FMS) as a single occurrence, avoiding the creation of "islands" of automation instead of creating an integrated production system for flexible production.

TECHNOLOGY AUDITS

In order to plan production innovation, it is important to have a complete list of all the technologies involved in production. Brian Twiss called a systematic identification and listing of all technologies in a firm a "technology audit" (Twiss, 1980). This is particularly useful for innovation in production systems because of the need to be systematic and avoid downtime in production. We will call a systematic list of all technologies involved in production a "production technology audit."

Hard-Good Production Technology Audit

A production technology audit should be oriented to the value-adding transformations of the firm producing the product, and may be extended "downstream" and "upstream" in the "industrial food chains" of the firm's businesses. The technologies for production of hard goods include:

1. Unit production processes
2. Materials handling processes
3. Production control processes
4. Environmental control processes
5. Product distribution processes

In each of these processes, there will be (1) physical equipment and phenomena and (2) control equipment and algorithms.

Accordingly, one can categorize a production technology audit for a hard-good producer in the kind of matrix of Table 18.1.

Service Delivery Technology Audit

A form of value-adding industrial chain also exists for service as a type of production. It differs from the manufacturing value-adding industrial chain in having more parallelism and being relatively less serialized. Service industries require constructing some kind of physical infrastructure; once this is in place, then service delivery requires the coordination of several subchains of value-adding sectors. Technologies for a service firm therefore group into:

1. Technologies for devices used in service delivery
2. Technologies for supply and maintenance of devices used in service delivery
3. Technologies for service delivery
4. Technologies for service development
5. Technologies for assisting the customer's applications of services
6. Technologies for communicating and conducting transactions with customers and suppliers
7. Technologies for controlling the activities of the service firm

TABLE 18.1 Hard Good Production Technology Audit

	Physical Equipment and Phenomena	Control Equipment and Algorithms
Unit Production Processes		
Materials Handling Processes		
Production Control Processes		
Environmental Control Processes		
Product Distribution Processes		

TABLE 18.2   Service Technology Audit

	Physical Equipment and Phenomena	Control Equipment and Algorithms
Necessary Infrastructure		
Major Device		
System		
Facility System		
Sales System		
Parallel Operating Systems		
Professional		
Personnel System		
Support Systems		
Scheduling		
Systems		
Maintenance/		
Resource Systems		

A service-delivery technology audit can be categorized, as shown in Table 18.2, in terms of (1) the necessary infrastructure and (2) the parallel operating systems. And within these, as in the manufacturing technology audit, (1) physical equipment and phenomena and (2) control equipment and algorithms can be identified.

Case Study: Production Systems of the Airline Industry

Consider the airline industry as an example of a service industry. The physical infrastructure that must be constructed or purchased to provide airline service includes (1) airplane systems, (2) airport systems, and (3) travel sales systems. With this infrastructure in place, the delivery of air transport services requires several subsystems to operate in parallel coordination.

For the airplane system, in addition to the airplane as principal device, (1) flight crew systems, (2) baggage handling systems, (3) airplane fueling and maintenance systems, and (4) food-handling systems need to operate simultaneously.

For the airport system, (1) passenger handling systems, (2) baggage handling systems, (3) land transport systems, (4) flight control systems, (5) security systems, (6) safety/emergency systems, and (7) cleaning/ maintenance systems need to operate simultaneously.

For the travel sales system, (1) reservation and scheduling systems, (2) travel agent systems, and (3) hotel and ground transportation reservation systems need to operate simultaneously.

PRODUCTION QUALITY

There are several meanings to the term "quality" in business:

1. *Quality of product performance*—How well does the product do the job for the customer?
2. *Quality of product dependability*—Does it do the job dependably?
3. *Quality of product variability*—can the product be produced in volume without defective copies?

The qualities of product performance and dependability are primarily determined by the design of the product and secondarily by its production. In contrast, product variability is primarily determined by the production process and secondarily by the product's design. Therefore, product quality is dependent on both design and production, but product variability primarily on production. We will call this the quality of product variability "production quality." Technological innovation that improves production quality provides a very important and powerful competitive edge.

Quality in Hard-Good Production

Production quality for hard goods particularly requires that the repeated acts of production occur with every detail under tight control. This requires extraordinary attention to detail, which is a distinctive cultural feature of the manufacturing community. Moreover, in hard-good production, this attention to detail—or its lack—has major impacts on costs and sales.

Genichi Taguchi and Clausing have discussed the total costs of poor quality for a hard good:

> When a product fails, you must replace it or fix it. In either case, you must track it, transport it, and apologize for it. Losses will be much greater than the costs of manufacture, and none of this expense will necessarily recoup the loss to your reputation. . . .
>
> —(Taguchi and Clausing, 1990, p. 65)

Taguchi and Clausing also maintained that the quality of a hard-good product should be seen from the perspective of the customer. From the customer's perspective, the important aspect of quality in a product is whether it works immediately on taking it out of the box and into the field, after it has possibly been knocked, dropped, and chipped.

Accordingly, technological innovations that improve the accuracy, speed, and throughput of unit production processes can give a producer a competitive edge.

Quality in Service Delivery

Quality in service production is primarily dependent on four criteria: efficacy, speed, cost, and safety. In routinized service delivery quality, efficacy is usually better established than speed, cost, and safety; therefore, these latter factors of quality are more important competitive factors. In customized service delivery, efficacy will be more uncertain to dominate over speed, cost, and safety issues. For example, banking services will be competitive primarily over speed, cost, and safety of funds, whereas medical services will emphasize efficacy and safety over speed and cost.

Accordingly, technological innovations that improve efficacy in applications or over more applications, increase safety, improve the responsiveness and speed of service delivery, or reduce service cost can provide powerful competitive edges to service providers. For example, Richard B. Chase and Robert H. Hayes identified characteristics for what they considered "world class" service delivery, which included seeking a reputation of service excellence, being innovative in improving operations, and continually seeking challenges for improvement in service quality and cost (Chase and Hayes, 1991).

PRODUCTION LEARNING CURVE

When a new production system (hard good, software, or service) is innovated, the initial production of the system will always be more costly, of lower quality, and less safe than later production. Learning how to produce better is the common experience of successful producers; this has been expressed in a form called the "production learning curve," as illustrated in Figure 18.3a. Plotting the unit cost of production often shows large unit cost reductions over time, as the technologies and skills of production are improved.

There are several mathematical formulas that can be used to model production learning curves, and all result in a declining curve on semi-log scales. A recent summary of mathematical forms appears in Badiru (1992). While these different models are of intellectual interest, in practice, which model is used is less important than how the curve is used as a management tool. It is important to use the production learning curve as a planning tool to set *targets* for production improvement.

In starting up a new production process, the production team should set "target goals" for production improvement and then try to meet them, as shown in Figure 18.3b. Setting target goals for production improvement is particularly important when innovating a new high-tech product for which new production processes may also need to be innovated. New high-tech products soon meet price competition from competitors, and the innovator should not only be improving the product, but also moving down the produc-

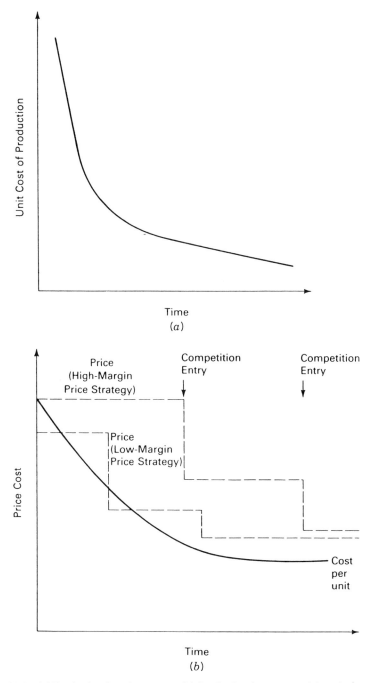

Figure 18.3. *(a)* Production learning curve. *(b)* Production improvement targets for a future production learning curve.

tion learning curve faster than competitors. This is one of the basic principles for a "first mover." Only maintaining a "High-Margin Price Strategy" simply invites competition to enter the market.

Expanding on the concept of the production-learning curve, Dorothy Leonard-Barton suggested that manufacturing be treated as a kind of learning laboratory emphasizing the production improvement processes—problem solving, continuous experimentation, and aggressive acquisition of new process knowledge (Leonard-Barton, 1992).

SUMMARY

Economies of scale in production are important to build when innovating new high-tech products, and economies of scope in production are important to build when product technologies mature.

Production is the organizational process that creates product volume; this has been called by different names in different industries. Production of hard goods and services differ markedly in their details, yet all production can be logically conceived of as a system of production composed of unit processes.

A periodic production technology audit, which tracks the number and kinds of technologies in a production, requires a systematic effort and can be useful to prepare for innovation in a production system.

The qualities of product performance and dependability are determined both by the design of the product and by its production. Product variability is an important perceived aspect of product quality by a customer. Setting target goals for production improvement is particularly important when innovating a new high-tech product for which new production processes may also need to be innovated.

FOR REFLECTION

Select an industry and plot the reduction of unit production costs for one of its product lines over time. What kinds of production technology innovations contributed to reduction of production costs and/or improvement of production quality?

___19

INTEGRATING TECHNOLOGY AND BUSINESS STRATEGY

CENTRAL CONCEPTS

1. Technology leader versus follower
2. Corporate conglomerates
3. Corporate diversification
4. Strategies for diversification
5. Business plan
6. Formal planning processes
7. Technology strategy
8. Chief technology officer
9. Technology acquisition

CASE STUDIES

3M Corporation's Technology Leadership Strategy for Diversification
The Demise of RCA
Formal Planning Process at Henkel KGaA in the 1980s

INTRODUCTION

We will now review how technology strategy, which underpins the future product and production plans of an enterprise, should and can be included in the business planning of a firm. We recall that technology strategy looks to the future technological capabilities of a firm, and, as we have seen, is complicated by (1) many kinds of technologies used by a business and (2) the number of different businesses in a diversified firm. Thus, the formulation of technology strategy can be complex, and its inclusion in business planning not simple. Yet it is very important to include technology strategy if a firm is to exploit its technical capabilities as one of its core competencies.

TECHNOLOGY LEADER VERSUS FOLLOWER

There are two general approaches in technology strategy: to be either a technology leader or a technology follower. Both strategies have proven successful under the right conditions.

The advantage of technology leadership is the opportunity to move first before competitors. First movers always have initial advantages to be the first to develop production capability, distribution capability, and brand recognition. The initial dominant market share in a new technology goes to first movers as technology leaders. Being first can also create production experience and capacity, which provide important lead times over competitors in cost, quality, and volume of product production. A first mover that continues to improve the product and lower production costs eventually emerges as the dominant firm in the maturing market of a new technology. Thus, the potential advantage to the technology leader is to create new markets and dominate in them.

However, there is a risk to being first. If the early product is not quite right for a market focus, then competitors can see the market more clearly than did the technology leader, who has shown it by pioneering. Then, if a competitor moves before the first mover can redesign the initial product and enters the market quickly with a refined product of superior performance and/or lower price, this competitor can take the market away from the first mover. Lower price, improved performance or features, and/or more focused marketing can give a technology follower a competitive advantage over a technology leader. Yet, to gain such a competitive advantage, the technology leader must make a mistake. Accordingly, the risk to the technology follower is that the technology leader might not make a mistake. When the technology leader makes no mistakes, the technology follower can never catch up or conquer.

Glen Urban and John Hauser have characterized firms' product development strategies as being proactive or reactive (Urban and Hauser, 1993). Proactive product strategies initiate product change before competitors, whereas reactive product strategies follow competitors' product innovations. Proactive product strategies arise from having research capability to create new technologies and engineering capabilities to design *new products embodying the new technologies*. Reactive product strategies can be *defensive* by protecting the profitability of existing products by introducing a redesigned product to counter a competitor's new product or be *imitative* by designing a "me-too" product to match a competitor's new product.

Both proactive and reactive product strategies can produce successful competitive positions, depending on how a firm benchmarks against competitors. When a firm leads competitors in research and engineering in a given area, proactive product strategies should be adopted. When a firm lags behind a competitor in research, reactive product strategies must be adopted. Accordingly, a firm's technology strategy determines the necessary proactive or reactive product strategies of its businesses.

Case Study: 3M Corporation's Technology Leadership Strategy for Diversification

This case study illustrates how a diversified corporation can use technology-leader strategy as a competitive strength. The historical context of this case study is in the second half of the twentieth century, when pressure was on large firms to generate growth. Most large firms after the industrial revolution grew as the industry upon which they were focused grew. Most companies were industrially focused, and most industries had core technologies. Thus, growth for a firm was tied to the market growth of the industry and the technological change in the core technologies of the industry. As the second half of the twentieth century began, however, many industries were entering maturity in their core technologies and saturation in their markets. The problem was how to maintain corporate growth outside the confines of single-industry companies. This issue stimulated the movement to corporate diversification.

The 3M Company is a large firm that has grown through internal diversification around its core technology strategies of coatings and adhesives. In 1980, 3M made 45,000 products (including product variations in size and shape). These were diverse products, from sandpaper and photocopiers to tape, skin lotions, and electrical connectors. The source of this product diversity was 3M's own innovations, with 95% of the company's sales ($5.4 billion in 1979) from products related to coating and bonding. This was 3M's secret for successful diversification—a core technology competency in material coatings and bonding technologies, which enabled the company to know the businesses it ran—for 3M had invented its own businesses. In 1980, there were 40 business divisions created from product innovations from 3M's research. Earnings for 3M had risen each year from 1951 through 1980 (except in 1972, the year of the oil crunch) (Smith, 1980).

The origin of 3M and its innovative culture can be traced back to its roots. Minnesota Mining & Manufacturing Co. began in 1902 at Two Harbors, Minnesota, when local investors purchased a mine. The mine was supposed to contain high-grade corundum, a hard mineral used in abrasives. Instead, the corundum was low-grade, useful only for ordinary sandpaper. Sandpaper, even then, was a commodity business with low profit margins. The disappointed investors decided to look for products with higher value.

The new company sent its sales personnel searching for innovative ideas. They went onto their customers' shop floors to look for needs for which no one was providing a product. In automobile factories, they saw workers choking on dust from the dry sandpaper. They reported this to 3M, and researchers created a sandpaper that could be used when wet. This was the first step in beginning 3M's technological capability in adhesives and coatings. It was also the first step in 3M's success formula—communication between salespeople and researchers for new product innovations.

The next product also came from observations by salespeople. They noticed that workers in automobile plants had a difficult time keeping paint from running on two-tone-painted cars. Richard Drew, a young technician in 3M's lab, invented masking tape. Drew next invented another major product. In 1930, he put adhesive on cellophane (cellophane itself had been invented at DuPont in 1924). John Borden (a sales manager at 3M) added a tape dispenser with a blade on it. Together, Drew's and Borden's product invention became the famous Scotch Tape® product.

This combination of research personnel and sales personnel jointly conceiving of new product innovations is a very powerful management mechanism—matching new technology with market need. Overall, about half of 3M's innovative products have come from 3M salespeople looking for needs in their customers' shops and offices (market pull); the other half has arisen from researchers' bright ideas, looking for applications (technology push).

As an example of the latter, 3M's research lab came up with a synthetic fabric from pressing rayon or nylon fibers together (unique in that it had no weave, like a felt material). The researchers thought first of using it for disposable diapers, but it was too expensive. Then they thought of using it for seamless cups for brassieres, which was also too expensive. The health care division then came up with the right application—surgical masks. These would be more comfortable to doctors and nurses than woven masks because they could be pressed into the right shape—and hospitals could afford them.

3M gives annual prizes—the Carlton Awards—to its best innovators. The award was named after Richard Carlton, who was president of 3M from 1949 to 1953. It is given annually to a few scientists who made major contributions to 3M's technology. For example, Paul Hansen developed a self-adhesive elastic bandage sold by 3M under the Coban label. Dennis Enright developed telephone-cable splicing connectors. Arthur Kotz developed an electronic system for microfilming directly from computers. W. H. Perlson's research in fluorine chemistry led to the development of agrichemical products for 3M and to Scotchgard® fabric-protection spray.

The strong tradition of innovation at 3M continued into the 1990s. For example, in 1992, Ronald A. Mitsch observed that the first ingredient in success for 3M was still to establish corporate strategies that exploit "technology competencies":

1. A corporate goal calls for 25% of sales to come from products created in the last five years.
2. The senior R&D executive reports to the CEO, although 3M is a highly diversified company with more than 50 different major operating units.
3. Employees' contributions to innovation are tangibly recognized and rewarded.
4. Technologies and innovation are pursued across division lines. (Mitsch, 1992, p.23)

3M's business strategy was to grow through technology leadership.

STRATEGY

The example of 3M's use of core technologies in adhesives and surfaces to create new business opportunities is a technology strategy. Let us review the general concept of strategy. In business, strategy is a process for setting the direction of the business's future. Some future will happen to the business with or without strategy. Moreover, the future of the business that the strategy envisions may not occur. Without strategy, however, the future that will occur to a business will have been missing the ability of management to prepare and influence that future. Strategy is, therefore, a management opportunity to help bring about a desired future. Strategy is direction; planning is how to go in that direction.

For example, Lowell Steele emphasized the difference between operational management and strategic management (Steele, 1989). Operational management is concerned with the day-to-day operations, producing and selling the products of the business. Strategic management is concerned with changing the procedures, processes, and products of the business to prepare for the future. Operational management values stasis and reproducibility; strategic management values foresight and change. Organizational survival depends on both kinds of management. Short-term competitiveness and profitability are determined by the efficiency of operational management, while long-term survival is determined by the effectiveness of strategic management.

Michael Porter has stated that operational effectiveness does not constitute strategy (Porter, 1990). He pointed out that even following all the "best management practices" of the day means only bringing a firm up to the productivity frontier of the time. As such, all companies that operate at the level of best practice do not necessarily distinguish themselves competitively from one another. Porter emphasized that strategies against competitors means going beyond operational effectiveness and best practices and differentiating the company: "The essence of strategy is choosing to perform activities differently than rivals do" (Porter, 1996, p. 64). This is why technology strategy is important to business strategy. Technology strategy provides an important means for differentiating products.

Strategic management must begin with a vision of where the company wants to go and how will it be differentiated from competitors. Business vision is a fundamental responsibility of the top management team. Without vision, any organization will continue to go in the same direction as current operational management is set up to perform. Robert Kaplan and David Norton, for example, have noted that implementing a vision requires a set of processes (Kaplan and Norton, 1996). The first process is translating the vision into a plan. The second process is communicating the plan and linking it to performance measures and the reward system. The third process involves setting business targets and allocating resources. The fourth process focuses on feedback from performance to the vision to refine or change the vision from the impacts of reality.

Case Study: The Demise of RCA

One of the reasons for the long-term failure of many large corporations has been operating without effective strategy. Business diversification without strategy has been one kind of frequent leadership failure.

This case study illustrates what happens to a firm that does not use strategy as a management tool. The historical context is the time in the second half of the twentieth century when many large firms diversified into different technology-based businesses without any attention to corporate core competencies. The life and death of RCA illustrates the final consequences of the absence of effective strategy.

The Radio Corporation of America began with the strategy of a high-tech and highly focused business. It was put together in the 1920s by David Sarnoff. To advance the then-new radio industry in the United States, Sarnoff organized a pooling of radio patents. His vision of RCA was as a high-tech consumer electronics firm. Under Sarnoff, RCA fulfilled much of this vision: innovating black and white television in the United States in the 1940s and innovating color television in the world in the 1960s.

However, the successors to David Sarnoff as CEO altered RCA's concept of its enterprise to make the company a diversified conglomerate. A firm that was once a technology leader became a group of businesses without focus or strategy. The troubles of RCA began during the 1960s, just, ironically, when the firm was innovating color television. RCA was expanded with the purchase of the National Broadcasting Corporation (the leading broadcasting network) and with the purchase of Hertz (the leading car-rental company) and many other totally unrelated businesses. The management idea was simple—growth by acquisition. The issues of whether or not management could properly manage the acquired businesses or whether the portfolio of acquired businesses acquired strengthened or weakened the whole corporation—were not considered.

The RCA leadership's management idea at that time was exceedingly simple—any large corporation could be managed by financial strategy alone. Business strategy, technology strategy, product strategy, manufacturing strategy, and marketing strategy seemed to be of little concern to the top management of RCA. It was indeed a simple idea. Could top management lead several diverse businesses of which they knew very little? From RCA's subsequent history, the answer was clear—it could not. From 1975 to 1981, RCA had many troubled acquisitions and poor corporate profitability. The state of the company was so bad that, during those six years, RCA had four CEOs (Nulty, 1981).

The third of these CEOs, Edgar Griffiths, found he had inherited a poor financial position at RCA. He instituted stronger financial controls, increased factory automation, decreased payroll, and began a process of divesture. He sold Random House books, an Alaskan telephone system, two food companies, an X-ray equipment firm, and firms producing aircraft radar and mobile radios. One can see from this list that RCA's leadership had no strategy other than financial strategy in acquiring firms— no complementary technology or business or market strategy. The company was just on a buying spree.

But did Griffiths himself learn any lessons from his predecessors' mistakes—any lessons about company strategy or culture? Apparently not, for Griffiths then went on his own buying spree. He bought more electronics businesses and purchased a financial firm, CIT Financial, for $1.4 billion in stock and cash (this being the biggest corporate purchase in 1980). These purchases doubled RCA's debt, and left it highly leveraged. Although RCA's earnings per share had increased from $2 to $3.50 between 1976 and 1981, Griffith's purchase of CIT tumbled the per-share earnings to negative 19 cents (Ehrbar, 1982).

RCA's successive leaders had fostered a corporate culture with a bad habit of taking heavy writeoffs. In 1971, RCA wrote off $490 million in the computer business. In 1981, RCA wrote off $230 million ($130 millon in TV picture-tubes, $59 million in truck leasing, and $34 million on TV shows). A fourth CEO, Thorton Bradshaw, succeeded Griffiths in 1981. He sold off CIT Financial, again taking a loss.

What was going on in this orgy of buying and selling of companies? In fact, the companies being bought and sold were not bad companies. Most were industrial leaders. For example, Hertz was acquired by RCA's Robert Sarnoff (who preceded Griffiths and had started the diversification strategy at RCA). At first, Hertz prospered under Robert Stone, assigned by Robert Sarnoff to manage Hertz. Hertz's profits increased fivefold, to $131 million, from 1971 to 1977. But RCA's leadership replaced Stone in 1977. Hertz's profits subsequently declined from 1979 through 1981. In 1985, RCA's leadership sold Hertz to United Airlines.

This example was illustrative of many of the other companies bought and sold by RCA. They did well before and after being owned by RCA, but not under RCA. RCA's leadership was not encouraging or rewarding good management who knew the businesses they were managing. As another example, Banquet Foods was profitable before and after RCA's ownership, but not during it. RCA's leadership under the fourth CEO, Bradshaw, seemed finally to be learning a lesson, however. Bradshaw was quoted as saying, "We didn't know how to run it [Banquet Foods], and we should not have had it" (Ehrbar, 1982, p. 67).

But, apparently, this lesson was never really learned by RCA. In December 1985, with RCA's share price depressed by a decade of mismanagement, General Electric Corporation acquired RCA and dismantled it. (The brand name of RCA Television was sold to the English electronics firm of Thompson Consumer Electronics.) After that, "RCA" still survived as a brand name, but not as a firm. One of the innovative high-technology firms of the early twentieth century, no longer existed by the end of the twentieth century, destroyed by management with a too-simple idea of an enterprise.

In this case, we can see how a succession of corporate leaders became wholly caught up in simple strategy—with the idea that only financial controls need be used to manage a company. RCA's leadership ignored the more complex issues of strategy, such as what businesses they should be in and how they could compete in those businesses. Companies are tough to kill, but it can be done. When a large company has poor leadership over a long period, neglecting any integrating strategies and trying to manage only by financial focus, the company can drift into the condition in which the stock market evaluates the whole of the firm as being less valuable than the sum of its parts. At this point, a corporate conglomerate becomes vulnerable for takeover and breakup. The Radio Corporation of America suffered such a succession of leadership failures that eventually led to its dissolution.

CORPORATE DIVERSIFICATION

Diversification of businesses within a firm is, in itself, not a bad idea. Properly done, diversification can spread the risk of survival across several businesses and strengthen the financial base of a firm. Most modern corporations are now diversified firms, containing different businesses. The desire to gain financial robustness from having many businesses is primary reason for diversification, which provides, to some degree, safety in recessions, escape from being dependent on a single low-growth market, and financial flexibility.

In addition, management also has reasons of self-interest for corporate growth through acquisitions. One motive is to acquire growth in order to maintain stock prices and prevent takeovers. Another reason is that, by increasing corporate size, management justifies increased management compensation.

CORE CORPORATE COMPETENCIES

There are strategies for corporate diversification that center on building a core competency, such as in the 3M illustration. Core competencies can center around several factors: technology, market, production, finance, or combinations of these.

Jon Didrichsen has noted that some corporations have shown broad technological competence in a scientific area, such as DuPont in chemicals or, as in the previous case study, 3M in coatings and adhesives (Didrichsen, 1972). In contrast, firms such as Procter & Gamble and Unilever have built businesses around market competencies in brands and distribution. C. K. Prahalad and Gary Hamel have argued that firms should identify core competencies (Prahalad and Hamel, 1990). More recently, Mark Gallon, Harold Stillman, and David Coates reiterated that core competencies can aggregate around technical or marketing competencies and added that these should provide competitive advantage through customer-perceived value, be difficult for competitors to imitate, and be extendable to new markets (Gallon et al., 1995).

The particular strength of diversification around core technical or market competencies is that businesses are built that the managers of the company understand and know how to run. Allan Kantrow pointed out that several factors should be included in diversification strategy:

> It is of great importance to identify and assess the nature of the relationship among a company's distinctive technological competence, its organizational structure, and its overall strategic orientation.
>
> —(Kantrow, 1980, p. 12)

Another observer, Richard Rosenbloom, also emphasized the importance of strategic thinking for linking technology and business strategies:

> The strategy framework is particularly appealing because it integrates in two relevant dimensions. First, the concept of strategy formulation calls for a perspective that cuts across the boundary of the organization, matching capability . . . with opportunity. . . . Second, the concept of strategy implementation re-

quires the translation of higher-level abstractions into more concrete and implementable terms.

—(Rosenbloom, 1978, p. 226)

In developing core technology competencies, the strategic branching of technologies can develop new product lines. For example, Dov Tzidony and Beno Zaidman suggest that core technology strategies can been generated from two goals:

1. To improve the chances for the long-term growth of a product line (by seeking technological innovation that improves the product to compete on technical performance or lower price)
2. To find new applications and/or customers for the product line (Tzidony and Zaidman, 1996)

BUSINESS AND TECHNOLOGY PLANNING

In 1996, a subcommittee of the Industrial Research Institute identified five "best practices" for improving the integration of technology planning with business planning:

1. Establish a structured process of technology planning.
2. Foster active involvement between R&D and other functional areas within the organization.
3. Get top management commitment to understand and support technology strategy.
4. Organize for effective technology planning and buy-in by all functions.
5. Hold both R&D and business units accountable for measurable results. (Metz, 1996)

It is important to have, throughout the company, a shared vision of what business success means for the company and R&D's role in that success. The formal means for doing this is to formulate and integrate technology strategy with business strategy.

FORMAL PLANNING PROCESSES

Let us briefly review the logic of any planning process. Planning is a process for envisioning a future and detailing a way to attain that future. Formal planning in organizations is important and necessary for:

1. Determining a direction for the organization's operations;
2. Communicating that direction to the participants in the organization in order to foster their cooperation in going in that direction; and
3. Making explicit the assumptions about the organization's environments in which the direction has been chosen.

The logic of planning requires several intellectual steps, which result in:

- Vision
- Goals
- Strategy
- Resources

All plans require a vision of a possible future. This vision sets the direction for the organization. Part of that vision must be the time period for the vision, the planning horizon—how long the plan is to be in effect and how far into the future the planning is to cover. Another part of the vision should be explicit assumptions about the environments in which the organization operates, and the changes and trends in these environments that the organization expects to encounter.

Next, a series of goals (or objectives) to be obtained in proceeding in the direction of the vision must be determined. The goals or objectives express the intention to follow the vision's direction, and set concrete outcomes to be attained in following it.

After vision and goals, a plan should next determine strategy. Strategy is the long-term manner in which the goals are to be attained, the means of going in the vision's direction to get to the concrete outcomes of the goals. Since strategy denotes long-term means, short-term tactics are also required—that is, the way things will be done in the near future to begin or evolve the strategic means.

The next part of a plan involves the resources required to perform the tactics and strategy in terms of facilities, equipment, and personnel. The last part of the plan is the budget—what it will cost to implement the tactics and strategy. The budget will be estimated annually and for the time period required to reach goals or to project to a planning horizon.

In summary, any formal plan should contain the following sections:

1. Vision—direction into the future
2. Planning horizon—distance into the future
3. Planning environments—trends in the contexts of the organization
4. Strategy—long-term means to attain the desired outcomes
5. Goals/objectives—concrete series of outcomes to be attained
6. Tactics—current and next-year means

7. Required resources—facilities, equipment, and personnel required for tactics and strategy
8. Budget—cost of resources

Not all organizations formally plan, but all organizations budget. Budgeting is the minimal level of planning. The reason there is often a lack of formal planning in many organizations is that they perform similar operations year after year. The concept of the enterprise pervades the organization, providing informal direction, and personnel carry on as they have always done. When no change is required in direction, such a situation may be adequate for a time. When change is required in the organization's direction or strategy, however, then formal planning is necessary.

In 1980, Frederick Gluck, Stephen Kaufman, and A. Steven Walleck published a study of the planning practices of 120 companies (Gluck et al., 1980). They found a range of degree of formalism of planning practiced by these companies:

1. Basic financial planning
2. Forecast-based planning
3. Externally oriented planning
4. Strategic planning

At the first level, only an annual budget is prepared in a functional format. The weakness of this planning is that the assumptions in preparing the budget are not explicitly spelled out. Consequently, management cannot compare the previous year's plan to current performance in order to judge whether management's assumptions about the concept of their enterprise are correct.

At the second level, forecast-based planning is usually begun when the need for future capital spending is recognized. Financial forecasts are then required to estimate the future return on investment of capital spending. The weakness of this form is that the assumptions on which forecasts are based are usually not explicit, nor are alternative assumptions tested.

At the third level, externally oriented planning, management sees past forecasts as inaccurate because the environments (business, economic, financial, regulatory, and technological) have changed and invalidated the past forecasts. Then, management tries to explicitly take into account the possibility of changes in the firm's environments in the forecasts and plans. However, externally oriented planning may not yet pursue the possibilities of how a firm may be strategically proactive in helping to shape changes in its environments.

At the fourth level, strategic planning, forecasts, environmental changes, and proactive strategies to deal with change are formulated. Technology strategy requires this level of corporate planning because technology strategy is a proactive plan to change the technical capabilities of the company, altering its environments and creating a desired technical and commercial future.

Case Study: Formal Planning Process at Henkel KGaA in the 1980s

This case study illustrates the fourth level of strategic planning in a diversified firm. The historical context of this case is the later part of the twentieth century, when large, diversified firms still struggled with the processes of formal planning.

In 1981, Henkel was organized by product areas, geographical areas, and functional areas, and with close attention to the market:

> Thinking and activity within the company is . . . strongly oriented to the careful observation of market activities.
> —(Grunewald and Vellmann, 1981, p. 20)

Henkel's planning process began a new annual cycle with the management board posing a new set of overall planning targets and by reviewing the corporate purpose. The targets used information from a forecast that extrapolated past performance into the future. The targets expressed desired levels of cash flow, return on investment, and levels of investment. The corporate purpose as a planning element was summarized in four interrelated areas: (1) fields of activity for the company with overall statements of the kinds of products, (2) technologies to be used, (3) consumer groups, and (4) geographic orientation.

The second step in Henkel's planning process was for corporate staff to take the targets and express them in a corporate-level strategic plan with an accompanying environmental analysis. This was provided to the firm's divisional and functional units.

The environmental analysis consisted of an economic forecast of the condition of the national economy (and other relevant international economies), including possible changes in government regulations. Baseline market forecasts were provided that estimated trends in the sales of product lines by application and customer. In addition, technological change was discussed. The planning environment linked three kinds of forecasts: economic, market, and technology. Henkel's environmental analysis emphasized the underlying factors affecting the company's businesses.

The "profit centers" of Henkel reviewed the corporate targets and planning environment in order to formulate profit-center-level goals and objectives. With these goals and objectives, Henkel's divisions then formulated tactics to reach the goals and budgets to fund the tactical activities. The individual plans of the profit-center-level activities were assembled into a total corporate plan and reviewed by Henkel's management board.

TECHNOLOGY STRATEGY

A technology strategy is an understanding and commitment to improving the knowledge and skill base of business practices. Technology strategies can be offensive or defensive. Offensive technology strategies aim to position a firm in a high-tech industry, and defensive technology strategies aim to defend a firm's position in a commodity industry. Both strategies are important, since a large, diversified firm will likely have both high-tech and commodity businesses.

Since an enterprise system is a value-adding transformation from resources into products or services, relevant technologies about which strategy to formulate should be determined by the technologies of relevant value-adding transformations for the firm. Changes in the technologies of resource acquisition will affect the value-adding activities of inbound logistics. Changes in the technologies of production processes will affect the value-adding activities of operations. Changes in the technologies of transportation and distribution will affect the value-adding activities of outbound logistics. Changes in the technologies of products will affect the value-adding activities of marketing, sales, and service. A company can be a high-tech company to the extent that it uses technological innovation in any of its enterprise subsystems to gain competitive advantages. Technology strategy should thus be formulated about improving all the knowledge bases of business systems:

- Product/service systems
- Production systems
- Distribution systems
- Information systems

Technologies in the industrial value chain of a business are also relevant to a business's technology strategy. A change in technology upstream will affect the types, quality, and cost of supplies for a business. A change in technology downstream of a business will affect the demand, quality, or price of a business's products or services. Michael Porter recommended formulating technology strategy for all the technologies of the industrial value chain:

1. Identify all the distinct technologies and subtechnologies in the firm and industrial value chains.
2. Identify potentially relevant technologies in other industries or under development from new science.
3. Determine the likely path of change of key technologies.

4. Determine which technologies and potential technological changes are most significant for competitive advantage and industry structure.
5. Assess a firm's relative capabilities in important technologies and the cost of making improvements.
6. Select a technology strategy, encompassing all important technologies, that reinforces the firm's overall competitive strategy.
7. Reinforce business unit technology strategies at the corporate level. (Porter, 1990)

The role of technology strategy in business planning should, therefore, identify the potential impact of technological change on any part of the relevant value chains: the firm's value chains, the encompassing industrial value chains, or adjacent and potentially substituting industrial value chains.

TECHNOLOGY PLANS

Technology plans anticipate and implement changes in the core and pacing technologies of an enterprise. Since technology plans need to be integrated into anticipated changes in the enterprise system, these plans need to detail anticipated impacts on:

- Enterprise evolution
- New or improved products or services
- New or improved production capabilities
- New or improved marketing capabilities
- Requirements for and impact on capitalization and asset capabilities
- New or improved organizational and operational capabilities

Accordingly, a complete business plan consists of:

1. Enterprise strategy
2. Product/service strategy
3. Manufacturing strategy
4. Marketing strategy
5. Financial strategy
6. Organizational strategy
7. Technology strategy

In the vision of the business plan, the core technologies of the firm should be delineated and their pace of change envisioned. For a future of rapid change in a core technology, the management team needs to commit to keeping the firm competitive in the core technology or decide to withdraw from businesses dependent on it.

In the planning environment of a business plan, it is important to forecast the directions and rates of change of the core technologies of the firm. It is also important to identify and forecast any potentially substituting technologies for any of the core technologies.

In the strategy of a business plan, it is necessary to formulate how to exploit technological change in the firm's businesses.

In the tactics of a business plan, each profit center must formulate how to exploit technological change in the improved or new product lines, services, or production.

In the required resources of a business plan, it is important to plan how research units of the firm need to contribute to divisional plans.

In the budgets of a business plan, R&D expenditures must be planned as part of the firm's investments in its future.

TECHNOLOGY SCENARIOS

In business planning, because the details of technological change can be obscure without an appropriate background, it is important for technical management to communicate technology's *impact* on business, rather than dwell on technical details. A useful way to do this is to discuss technological change in the business plan in the form of technological scenarios. Scenarios should be formulated in a "what-if" and forecasting mode, estimating, if certain technological progress were made, what the impacts would be on relevant product lines, production processes, services, or environmental conditions (Pyke, 1973).

Technology scenarios should help the corporation, during its strategic planning, focus on technological opportunities and their impact on market needs and business opportunities. Alan Kantrow succinctly expressed the technology-business connection:

> In short, what makes technology go is exactly what makes business go: coherent strategy and managers closely committed to it. To be effective, then, technological decisions must be strategically sound, for technology strategy and business strategy are of a piece.
>
> —(Kantrow, 1980, p. 20).

It can be useful to use metrics as a part of the technology strategy to numerically examine the contribution of technological innovation to the firm. However, the use of metrics as indicators of activities can be either helpful or misleading, depending how they are used for decision making.

Metrics alone never accurately measure complex activities, such as research, but they can be helpful. There are many metrics that can be used to help assess the value of R&D, some more or less useful. James W. Tripping, Eugene Zeffren, and Alan R. Fusfeld have made an extensive list of relevant metrics (Tripping, et al., 1995). These metrics include financial measures of return on research, projected value of R&D project portfolios, market shares of products or potential products, project development process time, and others. But metrics, however well used, cannot substitute for ideas. It is the direction of technological change—the new ideas for technology—that is far more important to technology strategy than metrics of past contributions of technology.

TECHNOLOGY IMPLEMENTATION PROCEDURES

Technology strategy results in both technology plans and technology implementation. The implementation of technology strategy occurs in procedures that integrate technology creation and development with commercial innovation of the technology in products, processes, or services. Procedures to facilitate this integration of implementation use logics of both technical and business development. For example, Peter R. Bridenbaugh described the logic of how the Alcoa Technical Center formalized connections between research and commercialization (Bridenbaugh, 1992). Figure 19.1 shows how the fundamental scientific understanding, core technology knowledge bases, and processes create concepts for commercialization that are demonstrated, scaled up, and implemented for commercialization. After commercialization, continuous improvement in a new product or production process

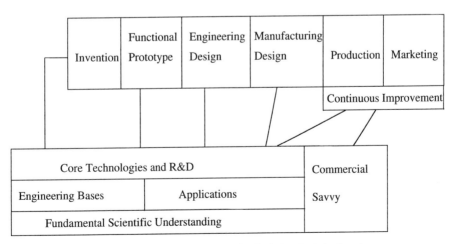

Figure 19.1. Technological innovation logic in a technical center.

feeds commercial savvy back to the core technology capabilities of the Technical Center.

CHIEF TECHNOLOGY OFFICER

In some firms in the 1990s, one means of trying to improve the integration of technology and business strategy is to create a position of chief technology officer (CTO). This title replaces the older title of head of the corporate laboratory or vice president of R&D. The new title entails broader responsibilities to reinforce the idea that technology and its implementation is the important focus of innovation, and not merely the upstream innovation stage of research and development.

James F. Mathis asserted that the role of the technology manager is conceptually broader than the role of the research manager, in that the CTO must see that the corporation has all the technologies it needs and in a timely manner for competitiveness (Mathis, 1992). As a later example, in 1996, Charles F. Larson noted that the role of R&D leadership in U.S. corporations had changed dramatically from the 1950s to the 1990s, with a new emphasis on the timeliness of research and speed in commercialization of new technology (Larson, 1996).

But a simple change in title is insufficient in reinforcing business experience in a technical career. For example, Bro Uttal, Alan Kantrow, Lawrence H. Linden, and B. Susan Stock examined 24 major U.S. corporations in 1992, finding that many gaps of credibility existed between R&D and line management (Uttal et al., 1992). Peter R. Bridenbaugh addressed this issue of credibility, pointing out that the credibility of the CTO needs to be built on several levels: personal, professional, and organizational. He argued that, in addition to technical acumen, CTOs must show commercial savvy, a global perspective, and the ability to form productive partnerships with line managers (Bridenbaugh, 1992). Peter Gwynne has pointed out that some U.S. firms have tried an organizational solution to the credibility problem of the CTO:

> Chief technology officers, who aim to deliver the value of technology for their companies, . . . often lack credibility. . . . Some companies have set out to remedy that by adding a line job to the normal duties of their CTOs.
> —(Gwynne, 1996, p. 14)

For example, some firms have added the responsibility for worldwide purchasing, or even managing a particular business of the firm. The combination of staff and line responsibilities also carries problems, however. The particular line responsibility can skew the focus of attention on the broader staff responsibility. An alternate solution might be to rotate researchers through line management jobs, so that in their total careers they have both staff and line experiences when they may become CTOs.

TECHNOLOGY ACQUISITION AND STRATEGIC RESEARCH PARTNERSHIPS

Acquisition of technology from outside the firm is an important source of technology in addition to technology created inside the firm. However, external acquisition of technology takes just as much planning and care as internal development.

For example, Nat C. Robertson discussed how Marion Laboratories acquired all its core technology from external sources during the 30 years of its existence (Robertson, 1992). Marion's first product, Os-Cal®, a calcium dietary supplement based on oyster shells, was based on technology acquired from outside, as have all its subsequent products. Marion has had many joint research projects with universities to develop the technologies for new products. The lesson that Marion Laboratories learned about technology acquisition is that its principal advantage is to reduce new product development time, but that the risks are great and the failure rate high. Organizing and staffing technology acquisition projects must be carefully planned and executed to reduce risk.

Strategic technology alliances between firms and universities and between firms and other firms is an important asset and problem in integrating technology and business strategies. Strategic research projects at the global level have become particularly important parts of many corporate strategies.

For example, in 1993, Toshiba had more than two dozen joint projects with different firms in Japan, the United States, Germany, South Korea, Canada, Italy, and Sweden. A major reason why Toshiba has entered into strategic research alliances is speed—reduced time to market: "[President Sato of Toshiba] thinks that carefully chosen partners are the key to moving quickly and marshaling the resources necessary to keep up with the high-tech race" (Kano, 1993, p. 120).

SUMMARY

The purpose of technology strategy is to improve the knowledge bases of business practice and establish a commitment to do so. Thus, the role of technology strategy in business planning is to include the foresight of technological change into the business vision in order to utilize technology as a competitive edge. A business can choose to be a technology leader or a technology follower, and either strategy can be successful under the right conditions. Business diversification developed on the basis of core competencies, including core technology and market competencies, facilitates creating businesses that corporate management knows how to run. This is the most robust strategy for growth that ensures long-term survival.

Formal planning in organizations determines a direction for the organization's operations, communicates that direction to the participants in the organization, and makes explicit the assumptions about the organization's environments in which the direction has been chosen. Technology planning enters the business plan directly in its vision, planning environment, strategy, tactics, resources, and budget sections.

FOR REFLECTION

Select a conglomerate and write a case history of its origin and performance. Has the conglomerate focused on any integrating competencies? What has been its stock price history? Has it been a target of any hostile takeovers? Were there any core competencies developed in the conglomerate?

20

ENVIRONMENT AND SUSTAINABLE ECONOMY

CENTRAL CONCEPTS

1. Green technology strategies
2. Ecological footprint
3. Pollution equals industrial inefficiency
4. Environmental analysis
5. Proper forms of government environmental regulation
6. Technology and societal wealth

CASE STUDY

Monsanto's Strategy for Sustainable Economy

INTRODUCTION

We conclude this overview of MOT by looking at the long-term impact of technology and economy on the environment. This is a very important problem, since, as we recall, the physical basis of economy comes from nature. As matter is conserved in the universe (except when it is converted to energy), materials that are taken from the environment are eventually returned in some form, to the environment. The processes of extracting energy from nature also alter the states of the physical environment. If the industrialization of the world is to continue as a positive societal force, the creation of economies that preserve the environment should be one of the highest priorities for technological innovation in the coming century.

Early industrialization did not address the impact of industrializing economies on nature. The question now is: How sustainable by nature are the present forms of industrialization? We know that the extensive clearing of forests for agricultural land has affected soil conservation and the oxygen-carbon dioxide cycles of the earth. We know that the vast expansion of human population has produced massive species extinctions in modern times. We know that certain industrial and agricultural chemicals have devastating effects on living creatures and even on the ozone layers in the upper atmosphere of the earth. We know that extensive irrigation results in increasing salinity of the soil. We have seen radioactive contamination of regions due to nuclear accidents and irresponsible pollution.

A major challenge to the world's industrialization, as the third century of industrial revolutions continues, is to create naturally sustainable industrial economies. This will require new and improved product and production technologies. We will now review the issues of environment, economy, and technology.

Case Study: Monsanto's Strategy for Sustainable Economy

This case illustrates how the CEO's vision of the long-term future of a company is essential to orient the personnel of a firm toward restructuring business to include environmental concerns in strategy. The historical period is the 1990s, when global competition and markets were increasing the awareness of large corporations that the whole planet was their market—and their environmental concern. In addition, the severe, extensive, and irresponsible environmental damage that had been done in the failed Communist economies of the twentieth century highlighted the need for environmentally sustainable economies.

In 1997, Robert B. Shapiro was chairman and CEO of Monsanto Company, based in St. Louis, Missouri. Shapiro had taken the initiative to reorient Monsanto toward a future business in environmentally sustainable economies (Magretta, 1997). It was a major strategic reorientation for Monsanto, since much of its traditional chemical business was in agricultural chemicals.

For an example of economy and environment, Shapiro used current potato production practices in the United States. For farmers to grow an annual crop of potatoes from seed potatoes, they needed to use pesticides against insects and viruses that would damage the crop. To produce the quantity of pesticides used annually required:

1. Starting with 4 million pounds of raw materials and 1,500 barrels of oil, to produce
2. 3.8 million pounds of inert ingredients in the product and 1.2 mil-

lion pounds of active insecticide in the product (plus 2.5 million pounds of waste in producing the product), which provided

3. 5 million pounds of inert and active ingredients of formulated insecticide product (in 180,000 containers)

4. Requiring 150,000 gallons of fuel to distribute and apply the insecticide.

Of this 5 million pounds of insecticide, only 5% actually reached the target pests of the potato crop. The other 95% (4.75 million pounds) of insecticide uselessly impacted the environment. One of the major potato growing areas in the United States where these insecticides were being applied was in Colorado, used to control the Colorado potato beetle.

How could this kind of useless environmental impact be changed? This was Shapiro's charge to his colleagues in Monsanto—find alternative technologies that could be environmentally friendly. Fortunately, Monsanto's early investment in biotechnology in the 1980s could provide the technology basis for a whole new approach. For example, Monsanto's scientists had used biotechnology to develop what Monsanto called its "NewLeaf" potato. This potato had been bioengineered to defend itself against the destructive Colorado potato beetle, and in 1995 it was already being used on potato farms in Colorado. Adding to the NewLeaf potato's resistance, Monsanto was engineered it to also include resistance to leaf virus.

Thus, Shapiro's positive attitude toward environmental challenges was very practical, based on new business opportunities that a new technology, biotechnology, was making possible:

> We can genetically code a plant, for example, to repel or destroy harmful insects. That means we don't have to spray the plant with pesticides. . . . If we put the right information in the plant, we waste less stuff and increase productivity. With biotechnology, we can accomplish that.
> —(Magretta, 1997, p. 82).

Monsanto's long-term agricultural business strategy was to move from being primarily a supplier of pesticides to being a supplier of biotechnologically engineered seeds and seed plants. The capability for Monsanto to see new business opportunities in a future of environmentally sustainable economy was due a succession of visionary business leaders who thought strategically about the future. As Shapiro commented:

> My predecessor, Dick Mahoney, understood that the way we [Monsanto] were doing things had to change. . . . Dick got way out ahead of the traditional culture in Monsanto and in the rest of the chemical industry. He set incredibly aggressive quantitative targets and deadlines. . . . In six years,

we reduced our toxic air emissions by 90%. . . . Dick put us on the right path.

—(Magretta, 1997, p. 84)

When Richard (Dick) Mahoney became president of Monsanto on September 1, 1983, Monsanto was a $6-billion-dollar company. Mahoney formulated new strategy to (1) reduce costs in Monsanto's traditional chemical-based businesses, and (2) position it in the new biotechnology and grow new businesses from it (Labich, 1984).

Mahoney was foreseeing major changes in the chemical industry. In the early 1980s, the world's chemical industry had been facing a slowdown in technological innovation in its traditional businesses (of explosives, dyes, fertilizers, and plastics). A new set of competitors on the horizon were in Middle Eastern countries, then making investments for future basic-chemicals businesses. The oil-producing companies in these countries were expected to bring new chemical plants on line with the cost advantage of their enormous oil and gas resources. Mahoney began formulating new technology strategy for Monsanto through biotechnology as a source of new high-tech products. In 1983, a technical journal, *Chemical & Engineering News* (C&EN), published an interview with Mahoney:

> **C&EN:** Where do you believe Monsanto is going in the next decade?
> **Mahoney:** We've set out on a pretty clear course that will see us evolving over time into a company that will be, in some ways, like we are today, and, in other ways, quite different.
> —(*Chemical & Engineering News,* 1983, p. 10)

The traditional businesses of Monsanto had been in basic chemicals, whose technologies of production had been relatively stable for years with no new radical invention. Consequently, basic chemicals had become commodity-type products. Monsanto sold some of these traditional businesses that were not profitable in the United States and abroad (including some fiber operations, a petrochemical plant in Texas, nylon operations in Europe, and a subsidiary chemical and plastic firm in Spain). Monsanto then turned its technology strategy toward biotechnology:

> **Mahoney:** Probably the one [area] that gets most press at Monsanto is the biological component. Here I'm talking about our existing business in agriculture and a new business in nutrition chemicals, which we started a few years ago to move into animal health and nutrition, and what we hope will be a substantial effort in the human health care area.
> —(*C&EN,* 1983, p. 10)

Monsanto had begun a large R&D investment in biotechnology:

Mahoney: We have perhaps $100 million of the $300 million per year in discovery—basic research. And $200 million in one form or another of applied research. Of the biotechnology area, virtually all of it is a component of the $100 million (basic research)—say $30 million or thereabouts. So about one third of our basic science is going into biotechnology.

—(*C&EN*, 1983, p. 13)

It was the vision of this earlier CEO, Mahoney, that provided the knowledge base for the next CEO, Shapiro, to build on for Monsanto's strategic future:

Ultimately, we'd love to figure out how to replace chemical processing plants with fields of growing plants—literally, green plants capable of producing chemicals. We have some leads: we can already produce polymers in soybeans, for example. But I think a big commercial breakthrough is a long way off. . . . I am not one of those techno-utopians who just assume that technology is going to take care of everyone. But I don't see an alternative to giving it out best shot. . . . The market is going to want sustainable systems, and if Monsanto provides them, we will do quite well for ourselves and our shareholders.

—(Magretta, 1997, p. 84)

This vision of Monsanto's CEO illustrates the important contribution of that leadership can make toward a firm's long-term future:

Sustainable development is going to be one of the organizing principles around which Monsanto and a lot of other institutions will probably define themselves in the years to come."

—(Magretta, 1997, p. 84)

GREEN STRATEGY

The adjective "green" became popular in the 1990s to indicate business strategies that included environmental concerns, such as "green manufacturing." We could use this term to characterize business strategy refocused to include environmental concerns as "green business strategy" and technology strategies that include environmental concern as "green technology strategy." To create sustainable economies on earth in the long term, it will take a lot more green thinking in future industrialization strategies than in those strategies of the past. We all would wish our posterity to inherit an industrialized earth that we have also left green with forest, meadows, and farmlands. The idea of the need for long-term economic sustainability began to gain currency among business leaders after the 1992 United Nations "Earth Summit" in Rio de Janeiro. In the 1990s, there were about 5.8 billion people in the world, with a population growth that still projected that number doubling over the following 40 years. About half of the people lived in abject poverty, a bitter kind of subsistence living worse than preindustrial tribal conditions.

Stuart L. Hart emphasized three kinds of current economies coexisting in the world: market economy, survival economy, and a nature economy (Hart, 1997). The market economy is the world of commerce comprised of both developed and developing nations. Survival economies exist in developing nations in rural areas and on the outskirts of cities, with people struggling to grow food and find water and fuel. They participate in the local market economy in marginal ways. The nature economy is the natural systems and resources utilized by both market and survival economies and affected by them (this is another name for an "environmental system"). Thus, as was emphasized at the Earth Summit conference, a broader picture of the world economy must include understanding of the different economies coexisting in the world—market, survival and natural economies—and their interactions.

One of these interactions is the impact of the market economy on the natural economy. This has been called an "ecological footprint" (Meadows, 1996). For example, the ecological footprint of the United States compared to the Netherlands and to India showed that it took 12 acres of natural economy to supply an average U.S. citizen's needs, compared to eight acres for a citizen of the Netherlands and one acre for a citizen of India. The ecological footprint was one way of expressing that different nations, with their different standards of living and different uses of technologies, were using different amounts of the earth's resources.

Both market economies and survival economies are very hard on the earth's environment. For example, some 15% of the topsoil in the United States was lost during the 1970s and 1980s to erosion, and irrigation of land in the western United States was increasing the salinity of soil. In survival economies, cutting down of trees and shrubs for fuel had denuded the watershed, and overfishing of offshore areas had depopulated fish. In market economies, cutting down of forests for timber and agriculture was eliminating vast areas of forest in developing countries, and excess commercial fishing in the second half of the twentieth century was depopulating the world's fish resources. Some have estimated that 40% of the planet's net primary productivity was already being utilized. Moreover, the extinction of many diverse species has proceeded at a terrible rate in the twentieth century.

The driving factors of the impact of economies on the environment are both sociological and technological. The primary sociological factor is the explosive growth of the earth's population due to (1) reduction of infectious diseases and improved public health, (2) lengthening of the average life span due to improved medical practices, and (3) continuation of high birth rates. The population has dramatically increased as people have lived longer and reproductive rates have been unchanged from preindustrial levels. The primary technological factor is the development of industrial technology without regard to environmental impact.

INDUSTRIAL INEFFICIENCY

Since technological innovation can improve productivity or product quality, such innovations can also reduce the impacts of production and products on the environment. One principle for "green" technological innovation is that any pollution from industry equals industrial inefficiency. For example, in the previous case of Monsanto, to produce 5 million pounds of insecticide product, 2.5 million pounds of industrial waste were also generated. Any reduction in this industrial waste (potential pollution) would increase productivity.

A second principle of green technological innovation is that products can be redesigned to improve quality when environment is part of the concept of product quality. For example, in the case of Monsanto, redesigning the potato genetically as the "NewLeaf" potato had eliminated the need for insecticide against the Colorado potato beetle. This new potato was a higher-quality product than a potato lacking resistance to the potato beetle.

ENVIRONMENTAL ANALYSIS

Scientific techniques from the environmental sciences and biology provide many tools for analyzing the impacts of industrial, economic, and population activities on the environment. For example, material cycles such as carbon, oxygen, or nitrogen can be used to measure and predict the impact of economic activities on these cycles. Combinations of material and energy cycles in techniques such as biomass systems can also estimate the impact of energy use and extraction on biological resources.

Environmental analysis should be a routine component in formulating industrial technology strategy. The benefits can be of several kinds, including:

- Lowering production costs by reducing waste
- Lowering environmental cleanup costs by reducing pollution
- Increasing the quality of brand recognition by being good corporate citizens
- Increasing competitive advantage through higher-quality products
- Investing in conservation of long-term resources

NATURAL ECONOMY AND SOCIETAL NEEDS

The concept of "natural economy" recognizes that all societies require nature to provide their basic physical and biological needs. Basic societal needs include:

- Air
- Water
- Food
- Shelter
- Energy
- Materials
- Transportation
- Communication
- Health
- Education
- Recreation
- Defense
- Justice

This list shows all the kinds of needs for which different technologies are used for different human purposes. Accordingly, one can understand that there are no obsolete societal needs, only obsolete technologies and industries that provide these needs. Green technologies, industries, and economies provide the basic societal needs in ways that alter the environment less than polluting technologies, industries, and economies.

When a societal need can be better satisfied by a substituting superior technology, the industry for which the inferior technology is a core technology will become obsolete. Thus, the idea of better matching of natural economy to market economy is necessary if sustainable economy is to become possible in the world's future.

ENVIRONMENTAL REGULATION AND INDUSTRIAL PRODUCTIVITY

The form of governmental regulation can have important effects on how industry contributes to environmental improvement. Michael E. Porter and Claas van der Linde have argued that properly formulated environmental regulation can foster the kind of technological innovation that both improves economy and environment:

> The need for regulation to protect the environment gets widespread but grudging acceptance: widespread because everyone wants a livable planet, grudging

because of the lingering belief that environmental regulations erode competitiveness. . . . This static view of environmental regulation . . . is incorrect. . . . Properly designed environmental standards can trigger innovations. . . . Ultimately, enhanced resource productivity makes companies more competitive, not less.

—(Porter and van der Linde, 1995, p. 120)

Porter and van der Linde suggested several criteria for a proper form of environmental regulation that promotes positive and effective innovative responses by industry for green technologies:

1. Create pressure that motivates companies to innovate.
2. Allow time for phasing in innovations.
3. Educate companies about resource inefficiencies and potential technological improvements.
4. Support research leading to technologies that promote environmentally friendly products and production.
5. Level the economic playing field for commercial transitions to environmentally friendly technologies.

Bad environmental regulation may not substantially improve the environment. For example, in 1986, the federal government passed the Superfund law for cleanup of environmental pollution. However, in a 1992 study, the Rand Institute for Civil Justice learned that 88% of the money paid between 1986 and 1989 by insurers on Superfund claims went for legal and administrative costs and only 12% was used for actual site cleanups (Acton and Dixon, 1992).

ENVIRONMENT AND WEALTH

The ideas of wealth and natural economy are not necessarily fundamentally in conflict. In fact, real wealth in society should be one in which the environment is desirable. For example, the wealthy in the world have always purchased secondary residences in regions of the world that are environmentally natural and beautiful.

In a societal context, the ability of one individual to satisfy needs better than another individual is one form of what we call "wealth"; we call this form of wealth "consumption." The second form of "wealth" is the ability of certain individuals to stimulate other individuals to work for their purposes; we call this form of wealth "capital."

Wealth as consumption or as capital can be socially accounted for on a measurement scale called "money." Basically, all economy is a barter system, and market economies mediate bartering through money. Money is a kind of societal accounting system for wealth, which elicits human cooperation on the basis of a mediated exchange of utility. Money is the intermediate of an economic barter system, so that the immediate values in exchanges of utility can be deferred to later and different exchanges.

With this concept of "wealth" as consumption or as capital, we can list the ways in which technology can expand the wealth, or prosperity, of a society:

1. Prosperity of capability
2. Prosperity of resources
3. Prosperity of time
4. Prosperity of knowledge
5. Prosperity of environment

One kind of wealth for humanity that technology makes possible is functional capability. For example, in ancient times, distant travel was expensive, time-consuming, and dangerous. Today, with modern commercial jets, travel of ordinary citizens between Europe, Asia, Africa, and the Americas is relatively inexpensive, quick, safe, and commonplace. Thus, technology can increase societal wealth through new functionality.

A second kind of wealth for humanity that technology makes possible is the availability of resources. For example, in the ancient civilizations of the Greek city-states and the Roman empire, only the kings were allowed to wear purple clothing. This was because the purple dye came from marine animals and was economically rare. Similarly, in ancient China, the color yellow was restricted to the emperor or empress. After the invention of artificial dyes in Germany in the 1800s, any kind of color became available cheaply. Today, anyone can wear purple or yellow. Thus, technology can increase societal wealth by lowering the cost of resources.

A third kind of wealth made possible by technology is the time required to accomplish tasks. The economic term for this is "productivity," the amount of labor time required to produce a unit of economic good or service. For example, one impact of the early industrial revolution was to apply power to drive the spinning and weaving machinery required to produce cloth. The quantity of cloth output that an individual worker could produce was extraordinarily increased by the introduction of powered textile machinery. Over time, the cost of clothing decreased with respect to other goods and services because less labor was required to produce cloth. Thus, technology can increase societal wealth by improving productivity.

A fourth kind of wealth for humanity from technology is the increase of knowledge. Technology improves the range and accuracy of perception through the creation of research instrumentation. Technology improves the theory of knowledge through the computational capabilities to model or simulate the dynamics of systems. Technology can thus increase societal wealth by providing more knowledge and culture.

A fifth kind of wealth that technology can make possible is to improve the environment for sustaining the scope of life. For example, soil and forests can be preserved through proper technologies of cultivation, harvesting, and proper management for long-term sustainability. Regions of the world that are abundant in natural plant and animal life can be set aside as natural parks, economically sustainable through properly controlled tourism.

The problem of natural economy and market economy is, therefore, not one of inherent conflict, but a multidimensional value problem in strategy. Sustainable economy strategy should be formulated as a many-dimensional problem in seeking technologies that simultaneously improve societal capabilities, resources, productivity, knowledge, and environmental quality. This multidimensional strategy for green technological innovation is not impossible, just a greater challenge for management, government, and research leadership.

SUMMARY

An "ecological footprint" is a way of expressing how different nations with different standards of living and different levels of technologies use different amounts of the earth's resources. As industrialization in the world continues, market economies and survival economies exist side by side, each making harsh demands on the environment as a natural economy for their resources.

Two principles of "green" technological innovation are that (1) any pollution from industry equals industrial inefficiency; and (2) products can be redesigned to improve quality when environment is a part of the concept of product quality. Proper forms of governmental regulation for environment encourage industrial innovativeness for "green" forms of new technology.

Environmental analysis provides a broad range of powerful tools for understanding and projecting economic impacts on the environment. Sustainable economic strategy should be formulated as a many-dimensional problem in seeking technologies that simultaneously improve societal capabilities, resources, productivity, knowledge, and environmental quality.

FOR REFLECTION

Select an industry and list the core technologies in production. From a government source, or industrial association source, find the kinds and volumes of industrial waste in this industry. What kinds of technological innovations would be desirable to reduce waste, and how much savings could be created by reduced waste?

BIBLIOGRAPHY

Abelson, Philip H., 1996. "Pharmaceuticals Based on Biotechnology," *Science,* Vol. 273, 9 August, p. 719.

Abnernathy, William J., 1978. *The Productivity Dilemma.* Baltimore, MD: Johns Hopkins University Press.

Abnernathy, William J., and K. B. Clark, 1985. "Mapping the Winds of Creative Destruction," *Research Policy,* Vol. 14, No. 1, pp. 2–22.

Acton, Jan Paul, and Lloyd S. Dixon, 1992. "Superfund and Transaction Costs: the Experiences of Insurers and Very Large Industrial Firms," working paper, Santa Monica, CA: Rand Institute for Civil Justice.

Afuah, Allan N., and Nik Bahram, 1995. "The Hypercube of Innovation," *Research Policy,* Vol. 24, pp. 51–76.

Ahl, David H., 1984. "The First Decade of Personal Computing," *Creative Computing,* Vol. 10, No. 11, pp. 30–45.

Allen, Thomas J., and Ralph Katz, 1992. "Age, Education, and the Technical Ladder," *IEEE Trans. on Engineering Management,* Vol. 39, No. 3, August, pp. 237–245.

Amatucci, Francis, and John H. Grant, 1991. "Eight Strategic Decisions that Weakened Gulf Oil," *Long Range Planning,* Vol. 26, No. 1, pp. 98–110.

Armst, Catherine, Judith H. Dorbrzynski, and Bart Ziegler, 1993. "Faith in a Stranger," *Business Week,* April 5, pp. 18–21.

Armstrong, Larry, and Peter Elstrom, 1996. "Inside Apple's Boardroom Coup," *Business Week,* February 19, pp. 28–30.

Ayres, Robert U., 1990. "Technological Transformations and Long Waves: Parts I and II," *Technology Forecasting and Social Change,* Vol. 37, pp. 1–37 and Vol. 37, pp. 111–137.

Badiru Adedeji, 1992. "Computational Survey of Univariate and Multivariate

Learning Curve Models," *IEEE Trans. on Engineering Management,* Vol. 39, No. 2, May, pp. 176–188.

Bailetti, Antonio, and John R. Callahan, 1995. "Managing Consistency Between Product Development and Public Standards Evolution," *Research Policy,* Vol. 24, pp. 913–931.

Bailetti, Antonio J., John R. Callahan, and Pat DiPietro, 1994. "A Coordination Structure Approach to the Management of Projects," *IEEE Trans. on Engineering Management,* Vol. 41, No. 4, November, pp. 394–402.

Bard, Jonathan F., Annura deSilva, and Andre Bergevin, 1997. "Evaluation Simulation Software for Postal Service Use: Technique Versus Perception,' *IEEE Trans. on Engineering Management,* Vol. 44, No. 1, February, pp. 31–42.

Bardeen, John, 1984. "To a Solid State," *Science 84,* November, pp. 143–145.

Bartimo, Jim, 1984. " 'Smalltalk' with Alan Kay," *InfoWorld,* June 11, pp. 58–61.

Baugh, S. Gayle, and Ralph M. Roberts, 1994. "Professional and Organizational Commitment Among Engineers: Conflicting or Complementing?" *IEEE Trans. on Engineering Management,* Vol. 41, No. 2, May, pp. 108–114.

Bastable, Marshall J., 1992. "From Breechloaders to Monster Guns: Sir William Armstrong and the Invention of Modern Artillery, 1854–1880," *Technology and Culture,* Vol. 33, pp. 213–247.

Bell, James R., 1984. "Patent Guidelines for Research Managers," *IEEE Trans. on Engineering Management,* Vol. 31, pp. 102–104.

Beltramini, Richard, 1996. "Concurrent Engineering: Information Acquisition between High Technology Marketeers and R&D Engineers in New Product Development," *Int. J. Technology Management,* Vol. 11, Nos. 1/2, pp. 58–66.

Berger, A., 1975. "Factors Influencing the Locus of Innovation Activity Leading to Scientific Instrument and Plastics Innovation," unpublished S.M. Thesis, Cambridge, MA: MIT Sloan School of Management.

Berkowitz, Leonard, 1993. "Getting the Most from Your Patents," *Research-Technology Management,* March–April, pp. 26–30.

Betz, Frederick, 1993. *Strategic Technology Management.* New York, McGraw-Hill.

Betz, Frederick, 1997. "Industry/University Centres in the USA: Connecting Industry to Science," *Industry and Higher Education,* Fall pp. 349–354.

Betz, Frederick, and Ian I. Mitroff, 1974. "Representational Systems Theory," *Management Science,* May 1974, pp. 1242–1252.

Bosomworth, Charles E., and Burton H. Sage, Jr., 1995. "How 26 Companies Manage Their Central Research," *Research-Technology Management,* May–June, pp. 32–40.

Bowen, H. Kent, Kim B. Clark, Charles A. Holloway, and Steven C. Wheelwright, 1994. "Development Projects: The Engine of Renewal," *Harvard Business Review,* September–October, pp. 110–120.

Boyden, J., 1976. "A Study of the Innovation Process in the Plastics Additives Industry," unpublished S.M. Thesis, Cambridge, MA: MIT Sloan School of Management.

Boyer, Edward, 1983. "Turning Glass to Plastic to Gold," *Fortune,* April 4, pp. 172–176.

Bricklin, Dan, and Bob Frankston, 1984. "Visicalc '79," *Creative Computing,* Vol. 10, No. 11, pp. 122–123.

Bridenbaugh, Peter R., 1992. "Credibility between CEO and CTO—A CTO's Perspective," *Research-Technology Management,* November–December, pp. 27–33.

Broad, William J., 1997. "Incredible Shrinking Transistor Nears Its Ultimate Limit," *The New York Times,* February 4, pp. C1–C5.

Burns, Tom, and G. M. Stalker, 1961. *The Management of Innovation.* London: Social Science Paperbacks.

Burrows, Peter, 1996. "The Man in the Disk Driver's Seat," *Business Week,* March 18, pp. 71–73.

Cabral-Cardoso, Carlos, and Roy L. Payne, 1996. "Instrumental and Supportive Use of Formal Selection Methods in R&D Project Selection," *IEEE Trans. on Engineering Management,* Vol. 43, No. 4, November, pp. 402–410.

Chandler, Alfred, 1990. "The Enduring Logic of Industrial Success," *Harvard Business Review,* March–April, pp. 130–140.

Chandrasekaran, Rajiv, and Victoria Shannon, 1997. "Struggling Apple Gets Boost From Microsoft," *The Washington Post,* August 7, pp. A1–A6.

Chase, Richard B., and Robert H. Hayes, 1991. "Beefing Up Operations in Service Firms," *Sloan Management Review,* Fall, pp. 15–16.

Chemical & Engineering News, 1983. "Monsanto's Richard Mahoney: Ready to Take on the 1980s," September 26, pp. 10–13.

Chester, Arthur N., 1994. "Aligning Technology with Business Strategy," *Research-Technology Management,* January–February, pp. 25–32.

Chiesa, Vittorio, 1996. "Managing the Internationalization of R&D Activities," *IEEE Trans. on Engineering Management,* Vol. 43, No. 1, February, pp. 7–23.

Christensen, Clayton, and Richard S. Rosenbloom, 1995. *Research Policy,* Vol. 24, pp. 233–257.

Cocoran, Elizabeth, 1997. "After 18 Months, Apple CEO Resigns," *The Washington Post,* July 10, p. E1.

Conover, Lloyd H., 1984. "Discovering Tetracycline," *Research Management,* Vol. XXVII, No. 5, pp. 17–22.

Costello, D., 1983. "A Practical Approach to R&D Project Selection," *Technology Forecasting & Social Change,* Vol. 23.

Cusumano, Michael A., 1988. *The Japanese Automobile Industry: Technology and Management at Nissan and Toyota.* Cambridge, MA: Harvard University Press.

Cusumano, Michael A., 1989. "Manufacturing Innovation: Lessons from the Japanese Auto Industry," *Sloan Management Review,* Fall, pp. 29–40.

Cyret, Richard M., and Praveen Kumar, 1994. "Technology Management and the Future," *IEEE Trans. on Engineering Management,* Vol. 41, No. 4, November, pp. 333–334.

DeBresson, Chris, 1995. "Predicting the Most Likely Diffusion Sequence of a New Technology Through the Economy: The Case of Superconductivity," *Research Policy,* Vol. 24, pp. 685–705.

Didrichsen, Jon, 1972. "The Development of Diversified and Conglomerate Firms in the United States, 1920–1970," *Business History Review,* Vol. 46, Summer, p. 210.

Drucker, Peter F., 1980. *Managing in Turbulent Times.* New York: Harper & Row.

Dunning, John H., 1994. "Multinational Enterprises and the Globalization of Innovative Capacity," *Research Policy,* Vol. 23, pp. 67–88.

The Economist, 1997. "Venture Capitalists," January 25, pp. 20–22.

Edleheit, Lewis S., 1995. "Renewing the Corporate R&D Laboratory," *Research-Technology Management,* November–December, pp. 14–18.

Ehrbar, A. F., 1982. "Splitting Up RCA," *Fortune,* March 22, pp. 62–76.

Elstrom, Peter, 1996. "PDA May Always Mean 'Pretty Darn Average'," *Business Week,* June 24, p. 110.

Elzinga, D. Jack, Thomas Horak, Chung-Yee Lee and Charles Bruner, 1995. "Business Process Management: Survey and Methodology," *IEEE Trans. of Engineering Management,* Vol. 42, No. 2, May, pp. 119–128.

Ettlie, John E., and Henry W. Stoll, 1990. *Managing the Design-Manufacturing Process.* New York: McGraw-Hill.

Ford, David, and Chris Ryan, 1981. "Taking Technology to Market," *Harvard Business Review,* March–April, pp. 117–126.

Forrester, Jay W., 1961. *Industrial Dynamics.* Cambridge, MA: MIT Press.

Fortune, 1993. "Information Technology Special Report: How to Bolster the Bottom Line," Autumn, pp. 15–28.

Foster, Richard, Lawrence Linden, Roger Whiteley, and Alan Kantrow, 1985. "Improving the Return of R&D II," *Research Management,* Vol. XXVII, No. 2, pp. 13–22.

Freeman, Christopher, 1974. *The Economics of Industrial Innovation.* New York: Penguin Books.

Frohman, Alan L., 1982. "Technology as a Competitive Weapon," *Harvard Business Review,* January–February, pp. 97–104.

Frosch, Robert A., 1996. "The Customer for R&D is Always Wrong!" *Research-Technology Management,* Vol. 40, No. 3, August, pp. 224–236.

Fusfeld, Herbert I., 1995. "Industrial Research—Where It's Been, Where It's Going," *Research-Technology Management,* July–August, pp. 52–56.

Gallon, Mark R., Harold M. Stillman, and David Coates, 1995. "Putting Core Competency Thinking into Practice," *Research-Technology Management,* May–June, pp. 20–28.

Gates, William, 1984. "A Trend toward Softness," *Creative Computing,* Vol. 10, No. 11, pp. 121–122.

Gemmell, Gary, and David Wilemon, 1994. "The Hidden Side of Leadership in Technical Team Management," *Research-Technology Management,* November–December, pp. 25–32.

General Electric, 1980. Descriptive material on the Corporate Research Laboratories, Schenectady, N.Y.

Geppert, Linda, 1994. "Industrial R&D: The New Priorities," *IEEE Spectrum,* September, pp. 30–41.

Gluck, Frederick W., Stephen P. Kaufman, and A. Steven Walleck, 1980. "Strategic Management for Competitive Advantage," *Harvard Business Review,* July–August, pp. 154–161.

Gomory, Ralph E., and Roland W. Schmitt, 1988. "Science and Product," *Science,* Vol. 240, May 27, pp. 1131–1204.

Graham, Alan K., and Peter M. Senge, 1980. "A Long-Wave Hypothesis of Innovation," *Technological Forecasting and Social Change,* Vol. 17, August, pp. 283–312.

Gray, Steven, B., 1984. "The Early Days of Personal Computers," *Creative Computing,* Vol. 10, no. 11, pp. 6–14.

Gross, Neil, 1997. "Information Technology Annual Report," *Business Week,* June 23, pp. 72–120.

Grunewald, Hans-Gunter, and Karlheinz Vellmann, 1981. "Integrating Regional and Functional Plans at Henkel," *Long Range Planning,* Vol. 14, No. 2, pp. 19–28.

Gwynne, Peter, 1993. "Directing Technology in Asia's 'Dragons'," *Research-Technology Management,* March–April, pp. 12–15.

Gwynne, Peter, 1996. "The CTO as Line Manager," *Research-Technology Management,* March–April, pp. 14–18.

Haeffner, Erik, 1980. "Critical Activities of the Innovation Process," pp. 1129–144, in *Current Information,* B. A. Vedin, ed. Stockholm: Almqvist & Wiksell.

Hart, Stuart L., 1997. "Strategies for a Sustainable World," *Harvard Business Review,* January–February, pp. 67–76.

Harvey, Jean, Louis A. Lefebvre, and Elisabeth Lefebvre, 1992. "Exploring the Relationship between Productivity Problems and Technology Adoption in Small Manufacturing Firms," *IEEE Trans. on Engineering Management,* Vol. 39, No. 4, November, pp. 352–358.

Hayes, Robert, and David Garvin, 1982. "Managing As If Innovation Mattered," *Harvard Business Review,* May–June, pp. 70–79.

Henderson, Rebeca, 1995. "Of Life Cycles Real and Imaginary: The Unexpectedly Long Old Age of Optical Lithography," *Research Policy,* Vol. 24, pp. 631–643.

Herink, Ritchie, 1994. Private communication.

Hutcheson, P., A. W. Pearson, and D. F. Ball, 1996. "Sources of Technological Innovation in the Network of Companies Providing Chemical Process Plant and Equipment," *Research Policy,* Vol. 25, pp. 25–41.

IRI, 1994. "First Annual Industrial Research Institute R&D Survey," *Research-Technology Management,* January–February, pp. 18–24.

Jantsch, Erich, 1967. *Technological Forecasting in Perspective.* Paris: Organization for Economic Cooperation and Development.

Jonash, Ronald S., 1996. "Strategic Technology Leveraging: Making Outsourcing Work for You," *Research-Technology Management,* March–April, pp. 19–25.

Judson, Horace Freeland, 1979. *The Eighth Day of Creation.* New York: Simon and Schuster.

Kahn, Joseph, 1996. "Clipped Wings: McDonnell Douglas's High Hopes for China Never Really Soared," *The Wall Street Journal,* May 22, pp. A1–A14.

Kano, Cindy, 1993. "How Toshiba Makes Alliances Work," *Fortune,* October 4, pp. 116–120.

Kantrow, Alan M., 1980. "The Strategy-Technology Connection," *Harvard Business Review,* July–August, pp. 6–21.

Kaplan, Robert S., and David P. Norton. 1996. "Using the Balanced Scorecard as a Strategic Management System," *Harvard Business Review,* January–February, pp. 75–85.

Kerzner, Harold, 1984. *Project Management: A Systems Approach.* New York: Van Nostrand Reinhold.

Kimura, Tatsuya, and Makoto Tezuka, 1992. "Managing R&D at Nippon Steel," *Research-Technology Management,* March–April, pp. 21–25.

King, Nelson, and Ann Majchrzak, 1996. "Concurrent Engineering Tools: Are the Human Issues Being Ignored?" *IEEE Trans. on Engineering Management,* Vol. 42, No. 2, May, pp. 189–201.

Kirkpatrick, David, 1997. "Intel's Amazing Profit Machine," *Fortune,* February 17, pp. 60–72.

Klevorick, Alvin K., Richard C. Levin, Richard R. Nelson, and Sidney G. Winter, 1995. "On the Sources and Significance of Interindustry Differences in Technological Opportunities," *Research Policy,* Vol. 24, pp. 185–205.

Klimstra, Paul D., and Ann T. Raphael, 1992. "Integrating R&D and Business Strategy," *Research-Technology Management,* January–February, pp. 22–28.

Knight, K. E., 1963. "A Study of Technological Innovation: The Evolution of Digital Computers," unpublished Ph.D. Dissertation, Pittsburgh, PA: Carnegie Institute of Technology.

Kocaoglu, D. F., 1994. "Special Issue on 40 Years of Technology Management," *IEEE Trans. on Engineering Management,* Vol. 41, No. 4. November, pp. 329–330.

Kokubo, Atsuro, 1992. "Japanese Competitive Intelligence for R&D," *Research-Technology Management,* January–February, pp. 33–34.

Kotter, John P., 1995. "Leading Change: Why Transformation Efforts Fail," *Harvard Business Review,* March–April, pp. 59–67.

Kumar, Uma, and Vinod Kumar, 1992. "Technological Innovation Diffusion: the Proliferation of Substitution Models and Easing the User's Dilemma," *IEEE Trans. on Engineering Management,* Vol. 39, No. 2, May, pp. 158–168.

Kunnathur, Anand S., and P. S. Sundararaghavan, 1992. "Issues in FMS Installation: A Field Study and Analysis," *IEEE Trans. on Engineering Management,* Vol. 39, No. 4, November, pp. 370–377.

Kuwahara, Yutaka, and Yasutsugu Takeda, 1990. "A Managerial Approach to Research and Development Cost-Effectiveness Evaluation," *IEEE Trans. on Engineering Management,* Vol. 37, No. 2, May, pp. 134–138.

Labich, Kenneth, 1984. "Monsanto's Brave New World," *Fortune,* April 30, pp. 57–98.

Larson, Charles F., 1996. "Critical Success Factors for R&D Leaders," *Research-Technology Management,* November–December, pp. 19–21.

Layton, Christopher, 1972. *Ten Innovations.* New York: Crane, Russak.

Lee, Denis M. S., 1992. "Job Challenge, Work Effort, and Job Performance of Young Engineers: A Causal Analysis," *IEEE Trans. on Engineering Management,* Vol. 39, No. 3, August, pp. 214–226.

Leonard-Barton, Dorothy, 1992. "The Factory as a Learning Laboratory," *Sloan Management Review,* Fall, pp. 23–38.

Lohr, Steve, 1997. "Creating Jobs," *The New York Times Magazine,* January 12, p. 15–19.

Loomis, Carol J., 1993. "Dinosaurs," *Fortune,* May 3, pp. 35–42.

Losee, Stephanie, 1994. "How Compaq Keeps the Magic Going," *Fortune,* February 21, pp. 90–92.

MacCormack, Alan David, Lawrence James Newman III, and Donald B. Rosenfield, 1994. "The New Dynamics of Global Manufacturing Site Location," *Sloan Management Review,* Summer, pp. 69–80.

MacLachlan, Alexander, 1995. "Trusting Outsiders To Do Your Research: How Does Industry Learn To Do It?" *Research-Technology Management,* November–December, pp. 48–53.

Marcua, Regina Fazio, 1994. "The Right Way to Go Global: An Interview with Whirlpool CEO David Whitwam," *Harvard Business Review,* March–April, pp. 135–145.

Margretta, Joan, 1997. "Growth through Global Sustainability: An Interview with Monsanto's CEO, Robert B. Shapiro," *Harvard Business Review,* January–February, pp. 78–90.

Marquis, Donald G., 1960. "The Anatomy of Successful Innovations," *Innovation,* November. Reprintcd in *Readings in the Management of Innovation,* M. L. Tushman and W. L. Morre, eds. Marshfield, MA: Pitman, 1982.

Mathis, James, 1992. "Turning R&D Managers into Technology Managers," *Research-Technology Management,* January–February, pp. 35–38.

McKendrick, David, 1994. "Sources of Imitation: Improving Bank Process Capabilities," *Research Policy,* Vol. 24, pp. 783–802.

McHugh, Josh, 1997. "Laser Dudes," *Forbes,* February 24, pp. 154–155.

Meadows, Donella, 1996. "Our 'Footprints' Are Treading Too Much Earth," *Charleston* (S.C.) *Gazette,* April 1.

Mensch, Gerhard, 1979. *Stalemate in Technology.* Cambridge, MA: Ballinger.

Metz, Philip D., 1996. "Integrating Technology Planning with Business Planning," *Research-Technology Management,* May–June, pp. 19–22.

Meyer, Christopher, and Ronald E. Purser, 1993. "Six Steps to Becoming a Fast-Cycle-Time Competitor," *Research-Technology Management,* September–October, pp. 41–48.

Meyer, Marc H., and James M. Utterback, 1993. "The Product Family and the Dynamics of Core Capability," *Sloan Management Review,* Spring, pp. 29–48.

Meyer, M. H., and J. M. Utterback, 1995. "Product Development Cycle Time and Commercial Success," *IEEE Trans. on Engineering Management,* Vol. 42, No. 4, November, pp. 297–304.

Mims, Forrest M., 1984. "The Altair Story," *Creative Computing,* Vol. 10, No. 11, pp. 17–27.

MIT 50K, 1996. *1996–1997 MIT 50K Entrepreneurship Competition.* Cambridge, MA: Massachusetts Institute of Technology.

Mitsch, Roland A. 1992. "R&D at 3M: Continuing to Play a Big Role," *Research-Technology Management,* September–October, pp. 22–26.

Morone, Joseph, 1993. "Technology and Competitive Advantage—the Role of General Management," *Management,* Vol. 36, March–April, pp. 16–25.

National Academy of Engineering and National Research Council, 1983. *The Competitive Status of the U.S. Pharmaceutical Industry.* Washington, DC: National Academy Press.

National Research Council, 1987. *Management of Technology: The Hidden Competitive Advantage.* Washington, DC: National Academy Press.

National Research Council, 1991. *Manufacturing Studies Board: Improving Engineering Design.* Washington, DC: National Academy Press.

National Science Board, 1996. *Science & Engineering Indicators 1996.* Washington, DC: Superintendent of Documents, U.S. Government Printing Office.

New York Times. 1984. February 20, p. D1.

Nobeoka, Kentaro, and Michael A. Cusumano, 1995. "Multiproject Strategy, Design Transfer and Project Performance," *IEEE Trans. on Engineering Management,* Vol. 42, No. 4, November, pp. 397–409.

Norling, Parry M., 1996. "Network or Not Work: Harnessing Technology Networks in DuPont," *Research-Technology Management,* January–February, pp. 42–48.

Nulty, Peter, 1981. "A Peacemaker Comes to RCA," *Fortune,* May 4, pp. 140–153.

Olby, Robert, 1974. *The Path to the Double Helix.* London: Macmillan; Seattle, WA: University of Washington Press.

Osborne, Adam, and John Dvorak, 1984. "Hypergrowth, Adam Osborn's Upcoming Book Tells His Side of the Story," *InfoWorld,* July 9, July 16, and July 23.

Patel, Pari, 1996. "Are Large Firms Internationalizing the Generation of Technology?" *IEEE Trans. on Engineering Management,* Vol. 43, No. 1, February, pp. 41–47.

Peterson, Michael, 1991. "Thomas Edison, Failure," *American Heritage of Invention & Technology,* Winter, pp. 9–14.

Porter, Michael, 1995. *Competitive Advantage of Nations.* New York: McGraw-Hill.

Porter, Michael E., and Class van der Linde, 1995. "Green and Competitive: Ending the Stalemate," *Harvard Business Review,* September–October, pp. 120–134.

Portugal, Franklin H., and Jack S. Cohen, 1977. *A Century of DNA.* Cambridge, MA: MIT Press.

Prahalad, C. K., and Gary Hamel, 1990. "The Core Competence of the Corporation," *Harvard Business Review,* May–June, pp. 79–91.

Pratt, Stanley E., and Jane K. Morris, 1984. *Pratt's Guide to Venture Capital Sources.* Wellesley, MA: Venture Economics Inc.

Purdon, William B., 1996. "Increasing R&D Effectiveness: Researchers as Business People," *Research-Technology Management,* July–August, pp. 48–56.

Pyke, Donald L., 1973. "Mapping—A System Concept for Displaying Alternatives," pp. 81–91, in *A Guide to Practical Technological Forecasting,* eds., James Bright and Milton Schoeman, Englewood Cliffs, NJ: Prentice-Hall.

Quinn, James Brian, 1985. "Managing Innovation: Controlled Chaos," *Harvard Business Review,* May–June, pp. 73–84.

Quinn, James Brian, Jordan J. Baruch, and Karen Anne Zien, 1996. "Software-Based Innovation," *Sloan Management Review,* Summer, pp. 11–24.

Ray, George F., 1980. "Innovation and the Long Cycle," in *Current Innovation,* B. A. Vedin, ed. Stockholm: Almqvist & Wiksell.

Rebello, Kathy, Peter Burrows, and Ira Sager, 1996. "The Fall of an American Icon," *Business Week,* February 5, pp. 34–42.

Reid, J. J., 1985. "The Chip," *Science 85,* February, pp. 32–41.

Robb, Walter, 1992. "Don't Change the Engineers—Change the Process," *Research-Technology Management,* March–April, pp. 8–9.

Roberts, Edward B., 1991. "High Stakes for High-Tech Entrepreneurs: Understanding Venture Capital Decision Making," *Sloan Management Review,* Winter, pp. 9–20.

Robinson, Nat C., 1992. "Technology Acquisition for Corporate Growth," *Research-Technology Management,* March–April, pp. 26–30.

Robertson, Arthur L., 1984. "One Billion Transistors on a Chip?" *Science,* Vol. 223, January 20, pp. 267–268.

Rosenberg, Nathan, and Richard R. Nelson, 1994. "American Universities and Technical Advance in Industry," *Research Policy,* Vol. 23, pp. 323–348.

Rosenbloom, Richard S., 1978. "Technological Innovation In Firms and Industries: An Assessment of the State of the Art" in *Technological Innovation,* P. Kelly and M. Kranzberg, eds., San Francisco, CA: San Francisco Press.

Rosenthal, Stephen R., and Anil Khurana, 1997. "Integrating the Fuzzy Front End of New Product Development," *Sloan Management Review,* Winter, pp. 103–118.

Rowen, Robert B., 1990. "Software Project Management under Incomplete and Ambiguous Specifications," *IEEE Trans. on Engineering Management,* Vol. 37, No. 1, February, pp. 10–21.

Rubenstein, Albert H., 1989. *Managing Technology in the Decentralized Firm.* New York: John Wiley & Sons.

Ryans, John K., Jr., and William L. Shanklin, 1984. "Positioning and Selecting Target Markets," *Research Management,* Vol. XXVII, No. 5, pp. 28–32.

Sanderson, Susan, and Mustafa Uzumeri, 1995a. "A Framework for Model and Product Family Competition," *Research Policy,* Vol. 24, pp. 583–607.

Sanderson, Susan, and Mustafa Uzumeri, 1995b. "Managing Product Families: The Case of the Sony Walkman," *Research Policy,* Vol. 24, pp. 761–782.

SAPPHO, 1972. "Success and Failure in Industrial Innovation," 2/72 Report on Project Sappho, Centre for the Study of Industrial Innovation, 162 Regent St., London, W1R6DD.

Schmidt, Robert L., and James R. Freeland, 1994. "Recent Progress in Modeling R&D Project-Selection Processes," *IEEE Trans. on Engineering Management,* Vol. 39, No. 2, May, pp. 189–201.

Schmitt, Roland W., 1985. "Successful Corporate R&D," *Harvard Business Review,* May–June, pp. 124–129.

Schwartz, Evan I., 1993. "In Poughkeepsie, A Bitter Family Breakup," *Business Week,* April 5, p. 22.

Serapio, Manuel G., 1995. "Growth of Japan-U.S. Cross-Border Investments in the Electronics Industry," *Research-Technology Management,* November–December, pp. 42–47.

Service, Robert F., 1997. "Making Single Electrons Compute," *Science,* Vol. 275, 17 January, pp. 303–304.

Shackil, Albert F., 1981. "Design Case History: Wang's Word Processor," *IEEE Spectrum,* Vol. 18, No. 1, pp. 29–34.

Shanklin, William L., 1983. "Supply-Side Marketing Can Restore 'Yankee Ingenuity'," *Research Management,* May–June, pp. 20–25.

Shrayer, Michael, 1984. "Confessions of a Naked Programmer," *Creative Computing,* Vol. 10, No. 11, pp. 130–131.

Smith, J. J., et al., 1984. "Lessons from 10 Case Studies in Innovation," *Research Management,* Vol. XXII, No. 5, pp. 23–27.

Smith, Lee, 1980. "The Lures and Limits of Innovation," *Fortune,* October 20, pp. 84–94.

Southtown Economist, 1993. "Business," Section V. Wednesday, February 10, p. 1.

Starr, Leon, 1992. "R&D in an International Company," *Research-Technology Management,* January–February, pp. 29–32.

Steel, Lowell W., 1989. *Managing Technology: The Strategic View.* New York: McGraw-Hill.

Stevenson, H. H., and D. E. Gumpert, 1985. "The Heart of Entrepreneurship," *Harvard Business Review,* March–April, pp. 85–94.

Stewart, Thomas A., 1993. "The King is Dead," *Fortune,* January 11, pp. 34–40.

Suarez, Fernando F., Michael A. Cusumano, and Charles H. Fine, 1995. "An Empirical Study of Flexibility in Manufacturing," *Sloan Management Review,* Fall, pp. 25–32.

Sugiura, Hideo, 1990. "How Honda Localizes Its Global Strategy," *Sloan Management Review,* Fall, pp. 77–82.

Swamidass, Paul M., and M. Dayne Aldridge, 1966. "Ten Rules for Timely Task Completion in Cross-Functional Teams," *Research-Technology Management,* July–August, pp. 12–13.

Taguchi, Genich, and Don Clausing, 1990. "Robust Quality," *Harvard Business Review,* January–February, pp. 65–75.

Thayer, Ann M., 1996. "Market, Investor Attitudes Challenge Developers of Biopharmaceuticals," *Chemical & Engineering News,* August 12, pp. 13–21.

Thomas, Sandra M., Keiko Kimura, and Julian F. Burke, 1995. *Research Policy,* Vol. 24, pp. 645–663.

Thomas, Lewis, 1984. "Medicine's Second Revolution," *Science 84,* November, pp. 93–98.

Thorne, Barbara, 1997. "Overview of Biopharmaceutical Development at Bristol-Meyers Squibb Co.," Presentation at Biotechnology Process Engineering Center, MIT, February 25.

Titus, George J., 1994. "Forty-Year Evolution of Engineering Research: A Case

Study of DuPont's Engineering Research and Development," *IEEE Trans. on Engineering Management,* Vol. 41, No. 4, November, pp. 350–353.

Townes, Charles H., 1984. "Harnessing Light," *Science 84,* November, pp. 153–155.

Tripping, James W., Eugene Zeffren, and Alan R. Fusfeld, 1995. "Assessing the Value of Your Technology," *Research-Technology Management,* September–October, pp. 22–39.

Tully, Shawn, 1993. "The Real Key to Creating Wealth," *Fortune,* September 20, pp. 38–50.

Tushman, Michael and Philip Anderson, 1986. "Technological Discontinuities and Organizational Environments," *Administrative Science Quarterly,* Vol. 31, pp. 439–465.

Twiss, Brian, 1980. *Managing Technological Innovation.* Harlow, Essex, England: Longman Group.

Tyre, Marcie J. and Wanda J. Orlikowski, 1993. "Exploiting Opportunities of Technological Improvement in Organizations," *Sloan Management Review,* Fall, pp. 13–26.

Tzidony, Dov and Beno Zaidman, 1996. "Method for Identifying R&D-Based Strategic Opportunities in the Process Industries," *IEEE Trans. on Engineering Management,* Vol. 43, No. 4, November, pp. 351–355.

Uenohara, Michiyuki, 1991. "A Management View of Japanese Corporate R&D," *Research-Technology Management,* November–December, pp. 17–23.

Urban, Glen, and Hauser, John, 1993. *Design and Marketing of New Products,* 2nd ed. Englewood Cliffs, NJ: Prentice-Hall.

Uttal, Bro, 1981. "Xerox Xooms toward the Office of the Future," *Fortune,* May 18, pp. 105–106.

Uttal, Bro, 1983. "The Lab That Ran Away from Xerox," *Fortune,* September 5, pp. 97–102.

Uttal, Bro, Alan Kantrow, Lawrence H. Linden, and B. Susan Stock, 1992. "Building R&D Leadership and Credibility," *Research-Technology Management,* May–June, pp. 15–23.

Utterback, James, 1978. "Management of Technology," pp. 137–160, in *Studies in Operations Management,* Arnoldo C. Hax, ed., Amsterdam: North-Holland Publishing.

Utterback, James M., and Fernando F. Suarez, 1993. "Innovation, Competition, and Industry Structure," *Research Policy,* Vol. 22, pp. 1–21.

Verity, John W., Thane Person, Deidre Depke, and Evan I. Schwartz, 1991. "The New IBM," *Business Week,* December 16, pp. 112–118.

Vesper, Karl H., 1980. *New Venture Strategies.* Englewood Cliffs, NJ: Prentice-Hall.

von Braun, Christoph-Friedrich, 1990. "The Acceleration Trap," *Sloan Management Review,* Fall, pp. 49–58.

von Braun, Christoph-Friedrich, 1991. "The Acceleration Trap in the Real World," *Sloan Management Review,* Summer, pp. 43–52.

Von Hippel, Eric, 1976. "The Dominant Role or Users in the Scientific Instrumentation Innovation Process," *Research Policy,* Vol. 5, No. 3, pp. 212–239.

Von Hippel, Eric, 1982. "Appropriability of Innovation Benefit as a Predictor of the Source of Innovation," *Research Policy,* Vol. 11, No. 2, pp. 95–115.

Von Hippel, Eric, and Marcie J. Tyre, 1995. "How Learning by Doing is Done: Problem Identification in Novel Process Equipment," *Research Policy,* Vol. 24, pp. 1–12.

Ward, E. Peter, 1981. "Planning for Technological Innovation—Developing the Necessary Nerve," *Long Range Planning,* Vol. 14, April, pp. 59–71.

Wheelwright, Steven C., and W. Earl Sasser, Jr., 1989. "The New Product Development Map," *Harvard Business Review,* May–June, pp. 112–125.

Whiteley, Roger L., Alden S. Bean, and M. Jean Russo, 1994. "Meet Your Competition: Results from the 1994 IRI/CIMS Annual R&D Survey," *Research-Technology Management,* January–February, pp. 18–25.

Wilson, Donald K., Roland Mueser, and Joseph A. Raelin, 1994. *Research-Technology Management,* July–August, pp. 51–55.

Wise, G., 1980. "A New Role for Professional Scientists in Industry: Industrial Research at General Electric, 1900–1916." *Technology and Culture,* Vol. 21, pp. 408–415.

Wohleber, Curt, 1993. "Straight Up," *Invention & Technology,* Winter, pp. 26–38.

Wolff, Michael F., 1984. "William D. Coolidge: Shirt-Sleeves Manager," *IEEE Spectrum,* May, pp. 81–85.

Workman, John P., 1995. "Engineering's Interactions with Marketing Groups in an Engineering-Driven Organization," *IEEE Trans. on Engineering Management,* Vol. 42, No. 2, May, pp. 129–139.

Ypsilanti, Dimitri, 1985. "The Semiconductor Industry," *The OECD Observer,* No. 132, pp. 14–20.

Zachary, G. Pascal, 1995. "Vannevar Bush on the Engineer's Role." *IEEE Spectrum,* July, pp. 65–69.

Zirger, B. J., and Janet L. Hartley, 1996. "The Effect of Acceleration Techniques on Product Development Time," *IEEE Trans. on Engineering Management,* Vol. 43, No. 2, May, pp. 143–152.

INDEX